# The Physics of Stars

## The Manchester Physics Series

General Editors

**F. MANDL: R. J. ELLISON: D. J. SANDIFORD**

*Physics Department, Faculty of Science,*
*University of Manchester*


| | |
|---|---|
| **Properties of Matter:** | B. H. Flowers and E. Mendoza |
| **Optics:** *Second Edition* | F. G. Smith and J. H. Thomson |
| **Statistical Physics:** *Second Edition* | F. Mandl |
| **Electromagnetism:** *Second Edition* | I.S. Grant and W. R. Phillips |
| **Statistics:** | R. J. Barlow |
| **Solid State Physics:** *Second Edition* | J. R. Hook and H. E. Hall |
| **Quantum Mechanics:** | F. Mandl |
| **Particle Physics:** | B. R. Martin and G. Shaw |

# THE PHYSICS OF STARS

**A. C. Phillips**

*Department of Physics and Astronomy*
*The University of Manchester*

John Wiley & Sons

CHICHESTER NEW YORK BRISBANE TORONTO SINGAPORE

Baffins Lane, Chichester
West Sussex PO19 1UD, England
Telephone (+44) (243) 779777



*Other Wiley Editorial Offices*

John Wiley & Sons, Inc., 605 Third Avenue,
New York, NY 10158-0012, USA

Jacaranda Wiley Ltd, 33 Park Road, Milton,
Queensland 4064, Australia

John Wiley & Sons (Canada) Ltd, 22 Worcester Road,
Rexdale, Ontario M9W 1L1, Canada

John Wiley & Sons (SEA) Pte Ltd, 37 Jalan Pemimpin #05-04,
Block B, Union Industrial Building, Singapore 2057

***Library of Congress Cataloging-in-Publication Data:***

Phillips, A. C.
The physics of stars / A. C. Phillips.
p. cm. — (Manchester physics series)
Includes bibliographical references and index.
ISBN 0 471 94057 7 — ISBN 0 471 94155 7 (pbk.)
1. Stars. 2. Astrophysics. I. Title. II. Series.
QB801.P48 1994
523.8—dc20 93–2079
CIP

***British Library Cataloguing in Publication Data:***

A catalogue record for this book is available from the British Library

ISBN 0 471 94057 7 (cloth)
0 471 94155 7 (paper)

Typeset in 10/12pt Times from author's disks by Text Processing Department,
John Wiley & Sons Ltd, Chichester
Printed and bound in Great Britain by Biddles Ltd, Guildford, Surrey

# Contents

# Editor's preface to the Manchester Physics Series

The first book in the Manchester Physics Series was published in 1970, and other titles have been added since, with total sales world-wide of more than a quarter of a million copies in English language editions and in translation. We have been extremely encouraged by the response of readers, both colleagues and students. The books have been reprinted many times, and some of our titles have been rewritten as new editions in order to take into account feedback received from readers and to reflect the changing style and needs of undergraduate courses.

The Manchester Physics Series is a series of textbooks at undergraduate level. It grew out of our experience at Manchester University Physics Department, widely shared elsewhere, that many textbooks contain much more material than can be accommodated in a typical undergraduate course and that this material is only rarely so arranged as to allow the definition of a shorter self-contained course. In planning these books, we have had two objectives. One was to produce short books: so that lecturers should find them attractive for undergraduate courses; so that students should not be frightened off by their encyclopaedic size or their price. To achieve this, we have been very selective in the choice of topics, with the emphasis on the basic physics together with some instructive, stimulating and useful applications. Our second aim was to produce books which allow courses of different length and difficulty to be selected, with emphasis on different applications. To achieve such flexibility we have encouraged authors to use flow diagrams showing the logical connections between different chapters and to put some topics in starred sections. These cover more advanced and alternative material which is not required for the understanding of later parts of each volume. Although these books were conceived as a series, each of them is self-contained and can be used independently of the others. Several of them are suitable for wider use in other sciences. Each author's preface gives details about the level, prerequisites, etc., of his volume.

We are extremely grateful to the many students and colleagues, at Manchester and elsewhere, whose helpful criticisms and stimulating comments have led to many improvements. Our particular thanks go to the authors for all the work they have done, for the many new ideas they have contributed, and for discussing patiently, and often accepting, our many suggestions and requests. We would also like to thank the publishers, John Wiley & Sons, who have been most helpful.

F. MANDL
R. J. ELLISON
D. J. SANDIFORD

*January, 1987*

# Author's Preface

Astrophysics is of natural interest to students and provides an ideal framework for demonstrating the power and elegance of physics. It is not surprising, therefore, that astrophysics is playing an increasing part in physics education. Despite this, there is a shortage of suitable textbooks for advanced undergraduates and beginning graduate students. For the most part, existing books are either too elementary and descriptive, or too technical and encyclopaedic.

This book is based on lectures prepared for a one semester course on stars for final-year undergraduates at Manchester University. To a large extent, the selection of topics covered has been based on a personal judgement as to whether the topic is important and whether it is also interesting to understand in terms of basic physics. The book is unusual in two respects.

First, there is a strong emphasis on explaining the underlying fundamental physics. Second, simple theoretical models are used to illustrate clearly the connections between fundamental physics and stellar properties. The overall aim is a self-contained, concise explanation of some of the most interesting aspects of stellar structure, evolution and nucleosynthesis.

In organizing the material in this book, I have recognized that the reader's motivation to understand physics is enhanced if the astrophysical application is near at hand and that an understanding of astrophysics requires a clear and concise reminder of physical principles. Thus, I have attempted to maintain a balance between physics and astrophysics throughout.

The first chapter introduces basic astrophysical concepts using elementary physical ideas which should be familiar to students pursuing a course on stars. Subsequent chapters rely on more advanced physical ideas which are normally met in the latter part of an undergraduate course. These ideas are carefully explained before they are applied. The properties of matter and radiation are considered in Chapter 2, heat transfer in Chapter 3, thermonuclear fusion in Chapter 4, stellar structure in Chapter 5, and the end-points of stellar evolution, namely white dwarfs, neutron stars and black holes, in Chapter 6. At the end of each chapter there are a number of problems aimed at testing understanding and extending knowledge. Hints for the solution of these problems are given at the end of the book.

In preparing the manuscript I have consulted many books and articles on astrophysics, particularly those listed in the Bibliography. It is important to mention here a subset books and articles which have been particularly influential. My interest in stellar physics was initially stimulated many years ago by the deep insight and directness of the articles by Salpeter, Weisskopf and Nauerberg. I have learnt much from two superb books: *Black Holes, White Dwarfs, and Neutron Stars* by Shapiro and Teukolsky, and *Neutrino Astrophysics* by Bahcall. In addition, Clayton's elegant article on *Solar Structure Without Computers* had a strong influence in Chapter 5. I have also found very useful the wealth of detail in *Cauldrons in the Cosmos, Nuclear Astrophysics* by Rolfs and Rodney, and in *Astrophysics I, Stars* by Bowers and Deeming.

Finally, I would like to express my thanks to colleagues at Manchester University. First, Franz Kahn and Franz Mandl read the early, primitive draft of the book, and their envouragement and help led me to take the idea of writing this book seriously; in particular, Franz Mandl's advice as Editor of the Manchester Physics Series was invaluable. Second, Judith McGovern and Mike Birse were very patient with me when I sought their help after doing stupid things with the word processor.

*May, 1993* A. C. PHILLIPS

CHAPTER 1

# Basic concepts in astrophysics

The aim of this book is to explore the properties of stellar interiors and hence understand the structure and evolution of stars. This exercise is largely based on the application of thermal and nuclear physics to matter and radiation at high temperatures and pressures. However before developing and applying this physics it is useful to consider the subject as a whole using elementary physics. In this brief and rapid overview we shall introduce some concepts which are fundamental to stellar evolution, fix the order of magnitude of some important astrophysical quantities and identify the basic observational information on stars. Many of the topics mentioned are covered in more detail later in the book and in the references listed at the end of the book. We begin by considering the processes which produced the raw material used in the construction of the first stars.

## 1.1 BIG BANG NUCLEOSYNTHESIS

To a first approximation matter in the universe consists of hydrogen and helium, with a smidgen of heavier elements such as carbon, oxygen and iron. It is now recognized that the bulk of this helium was produced by nuclear reactions which occurred during the first few minutes of the universe, a process called primordial or big bang nucleosynthesis. We shall begin this introductory chapter by giving a very brief outline of big bang nucleosynthesis so that the reader is aware of the origin and nature of the raw material used in the construction of the first stars.

### A brief history of the universe

In order to understand the history of the universe it is necessary to account for two important facts regarding the present universe: firstly the universe is expanding in such a way that if we extrapolate back in time it appears that the universe had infinite density some 10 to 20 billion years ago. Secondly the whole of space is filled with a thermal radiation at a temperature of about 3 K, the cosmic microwave background radiation discovered by Penzias and Wilson in 1965. These facts are consistent with the idea that the universe began with a sudden decompression, a big bang.

The big bang is not a local phenomenon with matter being expelled in all directions from a point in space. The big bang happened simultaneously everywhere in space. Everywhere was a point at the time of the big bang if the universe is closed; i.e. a finite volume of space with no boundary. But if the universe is open, the big bang occurred all over a space of infinite extent. According to the standard model of the big bang, the universe developed along the following lines:

- Nanoseconds after the big bang the universe was filled with a gas of fundamental particles: quarks and antiquarks, leptons and antileptons, neutrinos and antineutrinos, and gluons and photons. When the temperature fell below $10^{14}$ K, the quarks, antiquarks and gluons disappeared, annihilating and transforming into less massive particles. Fortunately because the number of quarks slightly exceeded the number of antiquarks, a few quarks were left behind to form the protons and neutrons present in today's universe. The heavier leptons and antileptons were also annihilated as the temperature fell.
- In the interval between a millisecond to a second after the big bang the universe consisted of a gas of neutrons and protons, electrons and positrons, neutrinos and antineutrinos, and photons. As the temperature fell, the density of the universe became too low for the neutrinos to interact effectively with matter; this occurred when the temperature was about $10^{10}$ K. These non-interacting, decoupled neutrinos now form a universal gas which, because of the expansion of space, has cooled to a temperature of about 2 K. As yet it has not been possible to detect this universal background of neutrinos. Soon after the decoupling of the neutrinos, the annihilation of electron-positron pairs removed all of the positrons and most of the electrons.
- After 100 seconds, neutrons combined with protons to form light nuclei, ultimately leading to a universe in which approximately 75% of the mass consists of hydrogen and 25% is helium. We shall explain later how these percentages were determined by the ratio of neutrons to protons in the universe when the neutrinos decoupled.
- After 300,000 years the temperature fell to 4000 K, low enough for the formation of stable atoms. Hydrogen and helium nuclei combined with electrons to form neutral hydrogen and helium atoms. As a result, the photons in the universe ceased to interact strongly with matter; in other words, the

universe became transparent to electromagnetic radiation. This radiation, freed from interaction with matter at a temperature near 4000 K, has now cooled to a temperature of about 3 K because of the expansion of the space. It is the cosmic microwave background radiation which was first detected by Penzias and Wilson. This radiation is slightly warmer than the as yet undetected neutrino background at 2 K because, unlike neutrinos, photons were warmed by the heat generated by electron–positron annihilation in the early universe.
- The universe continued to expand and cool until it reached its present lumpy condition with most of the matter assembled in stars, galaxies and clusters of galaxies.

This history of the universe is summarized in Table 1.1.

TABLE 1.1 A history of the universe according to the big bang. As the universe cooled quarks produced protons and neutrons, protons and neutrons formed helium and other light nuclei, and then nuclei and electrons combined to form neutral atoms. This led to today's thermal universe in which matter is assembled in stars and galaxies with a universal background of photons and neutrinos at temperatures of about 3 and 2 K, respectively.

| Cosmic time | Temperature | Events |
|---|---|---|
| $t \approx 10^{-4}$ s | $kT \approx 10^2$ MeV | Quarks form neutrons and protons |
| $t \approx 1$ s | $kT \approx 1$ MeV | Neutrinos decouple |
| $t \approx 4$ s | $kT \approx 0.5$ MeV | Electron–positron annihilation |
| $t \approx 3$ min | $kT \approx 0.1$ MeV | Helium and other light nuclei formed |
| $t \approx 3 \times 10^5$ years | $kT \approx 0.3$ eV | Atoms formed and photons decouple |

### The synthesis of helium

We shall now focus on the processes which led to the formation of helium and other light atomic nuclei. To understand these processes we shall follow what happened to the gas of neutrons and protons as the universe expanded and cooled from around $10^{10}$ to $10^9$ K. At temperatures above $10^{10}$ K, any deuteron formed from a neutron–proton collision was quickly disrupted by a collision because the thermal energies involved often exceeded the 2.2 MeV binding energy of the deuteron. The only nuclei existing at these temperatures were single protons and neutrons.

In normal circumstances a neutron beta decays with a mean life of about 15 minutes to a proton, an electron and an anti-neutrino,

$$n \rightarrow p + e^- + \overline{\nu}_e.$$

However at high temperature and density, neutrons can be transformed to protons, and protons can be transformed to neutrons in collisions involving thermal neutrinos, anti-neutrinos, electrons and positrons. In particular, neutrons and protons in the early universe were continually transformed into one another by the reactions:

$$\nu_e + n \rightleftharpoons e^- + p \qquad \text{and} \qquad \overline{\nu}_e + p \rightleftharpoons e^+ + n. \tag{1.1}$$

Because neutrons are more massive than protons, more energy had to be borrowed from the gas to make a neutron than a proton. Hence the neutrons were outnumbered by the protons. Indeed, the ratio of neutrons to protons at equilibrium at temperature $T$ is given by a Boltzmann factor:

$$\frac{N_n}{N_p} = \exp[-\Delta m\, c^2/kT], \tag{1.2}$$

where $\Delta m$ is the neutron–proton mass difference, 1.3 MeV/$c^2$.

The Boltzmann factor in Eq. (1.2) implies that the neutron–proton ratio decreased rapidly as the expanding universe cooled. But as the temperature and density decreased the neutrino reactions (1.1) became less frequent, and neutrons and protons were transformed into one another at a slower rate. Eventually, the reaction rates became too slow to maintain thermodynamic equilibrium. The neutrino reactions fizzled out, and the numbers of neutrons and protons ceased to change rapidly. Calculations indicate that the neutron–proton ratio became almost frozen at a value of about 1/5 when the temperature was just below $10^{10}$ K. In fact, this ratio continued to decline slowly because neutrons are unstable; they beta-decay to protons with a mean life of about 15 minutes.

After a few minutes, when neutron decay had reduced the neutron–proton ratio to about 1/7, the universe was cool enough for a sequence of two-body reactions to construct bound states of neutrons and protons. At about $10^9$ K deuteron nuclei began to be present in significant amounts as neutron–proton radiative capture, $n + p \rightarrow d + \gamma$, competed successfully with deuteron photodisintegration, $\gamma + d \rightarrow n + p$. Capture of neutrons and protons by deuterons led to the formation of tritons and helium-3. These nuclei in turn captured protons and neutrons to form helium-4. Since helium-4 is by far the most stable nucleus in this region of the periodic table, nearly all the neutrons that existed when the temperature was $10^9$ K were converted into helium-4. Moreover, the absence of stable nuclei with mass 5 and 8 prevented the formation of more massive nuclei, apart from small amounts of lithium-7.

Thus big bang nucleosynthesis took a gas of neutrons and protons and made helium-4 and a smattering of other light nuclei, namely deuterons, helium-3 and lithium-7 nuclei. All the neutrons were used in this construction, but many of the protons were left over. In fact, the theory of big bang nucleosynthesis makes a clear-cut prediction for the abundance of helium-4, but the predictions for the other light nuclei are less certain, being dependent on the uncertain density of the universe; see, for example, Bernstein *et al.* (1989).

We can estimate the helium-4 abundance produced in the big bang by noting that it is determined by the neutron to proton ratio in the universe just before nucleosynthesis. Because this ratio was about 1/7 we shall focus on 2 neutrons and 14 protons. These formed a single helium-4 nucleus containing 2 neutrons and 2 protons, and there were 12 protons left over. Thus 16 atomic mass units of neutrons and protons produced one helium nucleus of mass 4. The fraction of the mass converted into helium was 4/16 or 25%.

Hence, big bang nucleosynthesis led to a universe in which about 25% of mass was helium. The remaining 75% of the mass was mostly hydrogen formed from the left-over protons. This material was the raw material for the first stars.

## 1.2 GRAVITATIONAL CONTRACTION

Gravity is the driving force behind stellar evolution. Most importantly it leads to the compression of matter and thence to the formation of stars. It leads to the conditions where nuclear forces play a constructive role in thermonuclear fusion. The transformation of hydrogen to helium in the hot compressed centres of stars is often followed by a further compression and the transformation of helium into more massive elements such as carbon, oxygen and iron, the star dust out of which we are all made.

In order to identify some simple and general features of gravitational contraction, we consider in Fig. 1.1 a spherical system of mass $M$ and radius $R$, in which the only forces acting are due to its self-gravity and the internal pressure. To keep the analysis as simple as possible we shall assume spherical symmetry and no rotational motion. The density and pressure at a distance $r$ from the centre of the system will be denoted by $\rho(r)$ and $P(r)$.

We begin by finding an expression for the acceleration of a mass element located at a distance $r$ from the centre. The matter enclosed by a spherical shell of radius $r$ has mass

$$m(r) = \int_0^r \rho(r')\, 4\pi r'^2\, \mathrm{d}r',$$

and acts as a gravitational mass situated at the centre giving rise to an inward gravitational acceleration equal to

$$g(r) = \frac{Gm(r)}{r^2}.$$

There is also, in general, a force arising from the pressure gradient. To find this we consider a small volume element located between radii $r$ and $r + \Delta r$, of cross-sectional area $\Delta A$ and volume $\Delta r\, \Delta A$, as illustrated in Fig. 1.1. A net force arises if the pressure on the outer surface of the volume is not equal to the pressure on the

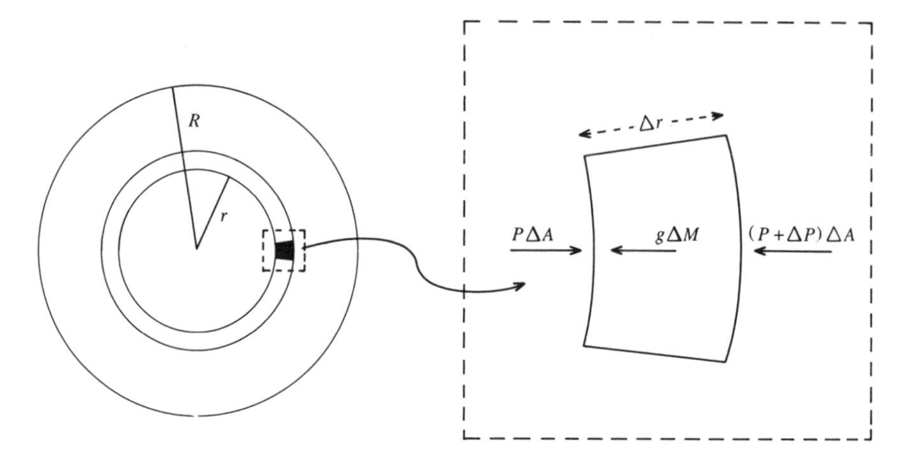


Fig. 1.1 A spherical system of mass $M$ and radius $R$. The forces acting on a small element with volume $\Delta r\ \Delta A$ at distance $r$ from the centre due to gravity and pressure are indicated. The gravitational attraction of the mass $m(r)$ within $r$ produces an inward force which is equal to $g(r)\ \rho(r)\ \Delta r\ \Delta A\ =\ g(r)\ \Delta M$. If there is a non-zero pressure gradient at $r$, the difference in pressure on the inner and outer surfaces leads to an additional force which can oppose gravity.

inner surface. Indeed, the inward force on the volume element due to the pressure gradient is

$$\left[P(r) + \frac{\mathrm{d}P}{\mathrm{d}r}\ \Delta r - P(r)\right]\ \Delta A = \frac{\mathrm{d}P}{\mathrm{d}r}\ \Delta r\ \Delta A.$$

Bearing in mind that the mass of the volume element is $\Delta M\ =\ \rho(r)\ \Delta r\ \Delta A$, we deduce that the inward acceleration of any element of mass at distance $r$ from the centre due to gravity and pressure is

$$-\frac{\mathrm{d}^2 r}{\mathrm{d}t^2} = g(r) + \frac{1}{\rho(r)}\ \frac{\mathrm{d}P}{\mathrm{d}r}. \tag{1.3}$$

Note that to oppose gravity the pressure must increase towards the centre.

## Free fall

We shall now assume that there is no pressure gradient to oppose gravitational collapse. In this case each mass element at $r$ moves towards the centre with an acceleration $g(r) = Gm(r)/r^2$. Spherical symmetry implies that each spherical shell of matter converges on the centre. In particular, a shell of matter enclosing a mass

$m_0$ collapses under gravity with an inward acceleration $Gm_0/r^2$, and the kinetic energy of the shell increases as its gravitational potential energy decreases. To find the inward velocity of the shell when its radius is $r$, we assume that the shell is initially at rest at a radius $r_0$, and that it encloses a mass which remains constant during collapse. The inward velocity can then be found from the conservation of energy equation:

$$\frac{1}{2}\left[\frac{\mathrm{d}r}{\mathrm{d}t}\right]^2 = \frac{Gm_0}{r} - \frac{Gm_0}{r_0}.$$

It follows that the time for free fall to the centre of the sphere is given by

$$t_{FF} = \int_{r_0}^{0} \frac{\mathrm{d}t}{\mathrm{d}r}\mathrm{d}r = -\int_{r_0}^{0} \left[\frac{2Gm_0}{r} - \frac{2Gm_0}{r_0}\right]^{-1/2} \mathrm{d}r.$$

This may be simplified by introducing the parameter $x = r/r_0$ to give

$$t_{FF} = \left[\frac{r_0^3}{2Gm_0}\right]^{1/2} \int_0^1 \left[\frac{x}{1-x}\right]^{1/2} \mathrm{d}x.$$

The integral in this equation may be evaluated by substitution of $x = \sin^2\theta$ to give $\pi/2$.

We have shown that the free-fall time for a shell of radius $r_0$ enclosing mass $m_0$ depends on $m_0/r_0^3$ ; i.e. it is determined by the average density of the matter enclosed. It follows that, in the absence of an internal pressure gradient, a sphere with an initial, uniform density of $\rho$ will collapse as a whole in a time given by

$$t_{FF} = \left[\frac{3\pi}{32G\rho}\right]^{1/2} \tag{1.4}$$

Collapse under gravity is never completely unopposed. In practice the energy released by the gravitational field of the collapsing system is usually dissipated into random thermal motion of the constituents, thereby creating a pressure which opposes further collapse. However, free fall is a relevant approximation if energy is easily lost by radiation, or if the constituents of the collapsing system can absorb energy by excitation or dissociation. For example, an interstellar cloud of molecular hydrogen can collapse rapidly as long as it is transparent to its own radiation, or as long as hydrogen molecules can be dissociated into atomic hydrogen, or as long as atomic hydrogen can be ionized. But the gravitational energy released in an opaque cloud of ionized hydrogen will be trapped as internal thermal motion. The internal pressure will rise and slow down the rate of collapse. The cloud will then approach hydrostatic equilibrium.

**Hydrostatic equilibrium**

Figure 1.1 and Eq. (1.3) indicate that an element of matter at a distance $r$ from the centre of a spherical system will be in hydrostatic equilibrium if the pressure gradient at $r$ is

$$\frac{\mathrm{d}P}{\mathrm{d}r} = -\frac{Gm(r)\rho(r)}{r^2}. \tag{1.5}$$

The whole system is in equilibrium if this equation is valid at all radii, $r$. In this case it is possible to derive a simple relation between the average internal pressure and the gravitational potential energy of the system.

To derive this relation we multiply Eq. (1.5) by $4\pi r^3$ and integrate from $r = 0$ to $r = R$ to obtain

$$\int_0^R 4\pi r^3 \frac{\mathrm{d}P}{\mathrm{d}r}\mathrm{d}r = -\int_0^R \frac{Gm(r)\rho(r)4\pi r^2}{r}\mathrm{d}r.$$

Both sides of this equation have simple physical significance. The right-hand side is simply the gravitational potential energy of the system:

$$E_{GR} = -\int_{m=0}^{m=M} \frac{Gm(r)}{r}\mathrm{d}m, \tag{1.6}$$

where $\mathrm{d}m$ is the mass between $r$ and $r + \mathrm{d}r$; i.e. $\rho(r)\ 4\pi r^2\ \mathrm{d}r$. The left-hand side can be integrated by parts to give

$$\left[P(r)4\pi r^3\right]_0^R - 3\int_0^R P(r)4\pi r^2 \mathrm{d}r.$$

The first term is zero because the pressure on the outside surface at $r = R$ is zero. The second term is equal to $-3\langle P\rangle V$, where $V$ is the volume of the system and $\langle P\rangle$ is the volume averaged pressure. Hence, we conclude that the average pressure needed to support a system with gravitational energy $E_{GR}$ and volume $V$ is given by

$$\langle P\rangle = -\frac{1}{3}\frac{E_{GR}}{V}. \tag{1.7}$$

In words, the average pressure is one third of the density of the stored gravitational energy. This expression for the average pressure needed to support a self-gravitating system is called the *Virial Theorem*.

The physical origin of this pressure depends on the system. Later in the book, in Chapter 2, we shall consider the pressures generated by classical and quantum gases of both non-relativistic and ultra-relativistic particles. But at this stage we would like to focus on the relation between the pressure and the internal energy density due to the translational motion of the particles, and in so doing we shall emphasize the profound difference in the behaviour of non-relativistic and ultra-relativistic systems.

To derive this relation we consider a gas of $N$ particles in a cubical box of volume $L^3$ with its edges orientated along the $x$, $y$ and $z$ axes. Initially we shall focus on a gas particle with velocity $\mathbf{v} = (v_x, v_y, v_z)$ and momentum $\mathbf{p} = (p_x, p_y, p_z)$. As this particle bounces around the box it strikes the sides at regular intervals. In particular, the rate at which it strikes one of the sides perpendicular to the $z$ axis is $v_z/2L$, and in so doing it imparts a momentum $2p_z$ with each strike. Hence the rate of momentum transfer to unit area of the side is $p_z v_z/L^3$. We now consider all $N$ particles in the box. The pressure due to these particles on a side perpendicular to the $z$ axis is

$$P = \frac{N}{L^3}\langle p_z v_z \rangle,$$

where the brackets denote an average over all the particles. If the gas is isotropic all directions of motion for the particles are equally likely and

$$\langle p_x v_x \rangle = \langle p_y v_y \rangle = \langle p_z v_z \rangle = \langle \mathbf{p} \cdot \mathbf{v} \rangle /3,$$

where

$$\mathbf{p} \cdot \mathbf{v} = p_x v_x + p_y v_y + p_z v_z.$$

Thus, the pressure on each side of the box is the same and equal to

$$P = \frac{n}{3}\langle \mathbf{p} \cdot \mathbf{v} \rangle, \tag{1.8}$$

where $n$ is the number of particles per unit volume.

Even though this expression for the pressure in an ideal gas has been derived using classical physics, it is also valid when quantum effects are important, as in a degenerate electron gas; see Section 2.1. Furthermore, it is also valid when the kinematics of the gas particles are described by special relativity.

We shall now compare and contrast two types of ideal gas, a gas of non-relativistic particles and a gas of ultra-relativistic particles. The general relation between the energy $\epsilon_p$ and the momentum $p$ of a particle of mass $m$ is

$$\epsilon_p^2 = p^2c^2 + m^2c^4,$$

and the velocity of the particle is $v = pc^2/\epsilon_p$. The familiar non-relativistic limit is found by assuming $p << mc$, so that $\epsilon_p = mc^2 + p^2/2m$ and $v = p/m$. The less familiar ultra-relativistic limit is found by assuming $p >> mc$, so that $\epsilon_p = pc$ and $v = c$. The general expression (1.8) for the pressure in an ideal gas takes the following forms in the the non-relativistic and ultra-relativistic limits:

- For a gas of non-relativistic particles of mass $m$, $\mathbf{p} \cdot \mathbf{v} = mv^2$ and the pressure becomes

$$P = \frac{2}{3}n\langle\frac{1}{2}mv^2\rangle = \frac{2}{3} \text{ of the translational kinetic energy density.} \tag{1.9}$$

- For a gas of ultra-relativistic particles $\mathbf{p}\cdot \mathbf{v} = pc$ and the pressure becomes

$$P = \frac{1}{3}n\langle pc\rangle = \frac{1}{3} \text{ of the translational kinetic energy density.} \tag{1.10}$$

We shall now show that the replacement of the factor of $\frac{2}{3}$ by $\frac{1}{3}$ when the particles become ultra-relativistic has a profound effect on the hydrostatic equilibrium of gases under gravity.

**Equilibrium of a gas of non-relativistic particles**

Consider a gas of volume $V$ held together by gravity. If the gas is ideal and if the gas particles are non-relativistic, then the average pressure implied by Eq. (1.9) is

$$\langle P\rangle = \frac{2}{3}\frac{E_{KE}}{V},$$

where $E_{KE}$ is the kinetic energy due to the translational motion of all the particles in the entire gas. Comparison with the average pressure needed for hydrostatic equilibrium, Eq. (1.7), shows that the gravitational and kinetic energies of an ideal gas of non-relativistic particles in hydrostatic equilibrium under their own gravity are related by

$$2E_{KE} + E_{GR} = 0. \tag{1.11}$$

If the particles have no internal excited degrees of freedom, the total energy of the gas is the sum of the kinetic and gravitational energies of the particles, $E_{TOT} = E_{KE} + E_{GR}$. Equation (1.11) implies that this total energy can be expressed in terms of either the kinetic energy or the gravitational energy of the particles; in particular

$$E_{TOT} = -E_{KE} \quad \text{and} \quad E_{TOT} = \frac{1}{2}E_{GR}. \tag{1.12}$$

Equations (1.11) and (1.12) are of fundamental importance in astrophysics. They can be used to describe the hydrostatic equilibrium of a system of self-gravitating non-relativistic particles.

The first point to note is that if such a system is in hydrostatic equilibrium, it is bound with a binding energy, $-E_{TOT}$, equal to the internal kinetic energy due to the translational motion of the gas particles. This implies that tightly bound clouds of gas have gas particles with high kinetic energy; in other words they are hot.

The second point to note is that if the system evolves slowly and remains close to hydrostatic equilibrium, the change in the gravitational and kinetic energies are simply related to the change in the total energy; for example, a 1% decrease in the total energy would be accompanied by a 2% decrease in the gravitational energy and a 1% increase in the kinetic energy.

Such changes characterize the behaviour of many astrophysical systems. For example let us consider a cloud of gas which is losing energy from its surface by radiation. If the energy loss from the surface of a gas cloud is supplied by the release of gravitational energy, the gravitational energy decreases and the internal thermal energy increases; the cloud will contract and get hotter. Indeed, for contraction close to hydrostatic equilibrium, half the gravitational energy released is lost from the surface and the other half is dissipated as heat; this heat provides the increase in pressure needed to oppose the increasing forces of gravity in the contracting cloud. However, if the energy loss from the surface can be supplied by the release of nuclear energy by thermonuclear fusion, the total energy $E_{KE} + E_{GR}$ remains constant and there is no need for the cloud to contract; the sun behaves in this way. But if nuclear fusion releases excess energy, there is an increase in the total energy. This implies an increase in the gravitational energy and a decrease in the kinetic energy; the cloud expands and cools. Conversely, nuclear reactions which absorb energy will cause a gas cloud to contract and heat up.

### Equilibrium of a gas of ultra-relativistic particles

We shall now show that the situation with regard to hydrostatic equilibrium is markedly different when a gas of ultra-relativistic particles is held together by gravity. In this case the pressure inside the gas is given by Eq. (1.10), and consequently the average pressure in the system is one third of the average kinetic energy density. If we equate this pressure to the average needed for hydrostatic equilibrium, Eq. (1.7), we find that the kinetic and gravitational energies are now related by

$$E_{KE} + E_{GR} = 0. \tag{1.13}$$

In words, hydrostatic equilibrium is possible only if the binding energy is zero. We have a system which is on the cusp of being bound and unbound. Indeed, as the ultra-relativistic limit is approached the binding energy decreases and the system is easily disrupted. This type of instability occurs in stars in which a substantial fraction of the pressure arises from radiation; i.e. from a gas of ultra-relativistic particles called photons. It can also occur in stars supported by the pressure of a gas of degenerate electrons if these electrons become very energetic. These instabilities are considered in detail in Sections 5.4 and 6.1.

### Equilibrium and the adiabatic index

The stability of hydrostatic equilibrium is often described in terms of the adiabatic index, $\gamma$, of the gas. This is particularly useful when the constituents of the gas have vibrational and rotational degrees of freedom.

The adiabatic index $\gamma$ is used to describe the relation between the pressure and the volume of a gas during an adiabatic compression or expansion. For such a process, $PV^\gamma$ is a constant; i.e. for small adiabatic changes in the volume and pressure

$$\gamma \frac{\mathrm{d}V}{V} + \frac{\mathrm{d}P}{P} = 0, \quad \text{or} \quad \mathrm{d}(PV) = P\,\mathrm{d}V + V\,\mathrm{d}P = -(\gamma - 1)P\,\mathrm{d}V.$$

As there is no heat transfer in an adiabatic compression or expansion, the change in the internal energy of the system is determined solely by the work done. If we denote the internal energy due to translational kinetic energy and the excited internal degrees of freedom of the gas particles by $E_{IN}$, then

$$\mathrm{d}E_{IN} = -P\,\mathrm{d}V,$$

and hence

$$\mathrm{d}E_{IN} = \frac{1}{\gamma - 1}\,\mathrm{d}(PV).$$

If the adiabatic index $\gamma$ is constant, we can deduce the following useful relation between the internal energy and the pressure of the gas:

$$E_{IN} = \frac{1}{(\gamma - 1)}\,PV.$$

We now consider a self-gravitating gas with adiabatic index $\gamma$, which is in hydrostatic equilibrium. The average pressure in such a gas can be expressed in terms of the internal energy and $\gamma$, and, by using the virial theorem Eq. (1.7), in terms of the gravitational potential energy:

$$\langle P \rangle = (\gamma - 1)\,\frac{E_{IN}}{V} = -\frac{1}{3}\,\frac{E_{GR}}{V}.$$

Thus, a self-gravitating gas with adiabatic index $\gamma$ is in hydrostatic equilibrium if

$$3(\gamma - 1)\, E_{IN} + E_{GR} = 0. \tag{1.14}$$

Equations (1.11) and (1.13) are particular cases of this more general relation between the internal and gravitational potential energies of a gas. These particular cases can be obtained from Eq. (1.14) by specializing to a gas of particles with no excited internal degrees of freedom so that $E_{IN} = E_{KE}$, the internal kinetic energy due to translational motion of the particles, and then setting $\gamma = 5/3$ for non-relativistic particles and $\gamma = 4/3$ for ultra-relativistic particles.

The total energy of a gas with adiabatic index $\gamma$ in hydrostatic equilibrium is given by

$$E_{TOT} = E_{IN} + E_{GR} = -(3\gamma - 4)\, E_{IN}. \tag{1.15}$$

We note that the gas is bound if $\gamma > 4/3$. Furthermore, the binding energy is small if $\gamma$ is near to 4/3, and when this is the case a small change in the total energy is accompanied by much larger changes in the internal and gravitational energies. For example, if $\gamma$ is 1% bigger than 4/3, a 1% decrease in the total energy is accompanied by a 25% increase in the internal kinetic energy and a 26% decrease in the gravitational potential energy. It is clear that the stability of such a system is precarious. Indeed, instability is expected whenever $\gamma$ is reduced towards 4/3. In this context, we note that for particles with no excited internal degrees of freedom $\gamma = 5/3$ when they are non-relativistic, but $\gamma$ approaches 4/3 as they become predominantly ultra-relativistic. The adiabatic index can also approach 4/3 when there are processes which provide new ways of absorbing heat, such as the dissociation of molecules, the ionization of atoms, the photodisintegration of atomic nuclei or the production of particles. Such processes will tend to render hydrostatic equilibrium precarious.

## 1.3 STAR FORMATION

It seems that most stars are formed in clusters. There are two characteristic kinds of clusters, globular and open. Globular clusters are compact aggregations of many thousands of stars. Studies of their spectra indicate that the member stars are deficient in heavy elements, such as carbon, oxygen and iron. This lack of heavy elements suggests that these stars are old stars formed from primordial hydrogen and helium. In contrast, open clusters are loose collections of 50 to 1000 stars. These stars are rich in heavy elements, indicating that they are comparatively young stars formed from matter which has been enriched with elements formed by earlier generations of stars.

There is as yet no complete understanding of how stars emerge from interstellar gas clouds. These clouds seem to have too much kinetic energy and too much angular momentum to condense into stars, and there is much interest in how this excess energy and angular momentum can be shed. Despite this uncertainty some general features of star formation can be identified. To do this we shall give a qualitative description of the gravitational contraction of clouds of uniform density.

**Conditions for gravitational collapse**

In order to begin the process of the condensation into a cluster of stars, a gas cloud must be sufficiently compact so that the attractive forces due to gravity are not overwhelmed by the dispersive effects of the internal pressure. In particular, the cloud becomes bound if the magnitude of the gravitational potential energy is larger than the internal kinetic energy. We shall determine an approximate condition for condensation by considering a cloud of radius $R$ and mass $M$ containing $N$ particles with average mass $\overline{m}$ at a uniform temperature $T$; for simplicity we shall assume that the cloud consists predominantly of hydrogen.

The gravitational potential energy can be evaluated with the aid of Eq. (1.6) to give

$$E_{GR} = -f\frac{GM^2}{R}, \tag{1.16}$$

where $f$ is a numerical factor which depends on the density distribution within the cloud. It is straightforward to show that $f = \frac{3}{5}$ for a spherical cloud of uniform density, but a larger value is obtained if the density is higher towards the centre; in our rough calculation we shall adopt a value of unity for $f$. The thermal kinetic energy of the cloud is found by noting that each particle contributes $\frac{3}{2}kT$. Hence

$$E_{KE} = \frac{3}{2}NkT. \tag{1.17}$$

The critical condition for the onset of condensation is

$$|E_{GR}| > E_{IN}. \tag{1.18}$$

This condition implies that a cloud of radius $R$ can condense if its mass exceeds

$$M_J = \frac{3kT}{2G\overline{m}}R.$$

It also implies that a cloud of mass $M$ can condense if its average density exceeds

$$\rho_J = \frac{3}{4\pi M^2}\left[\frac{3kT}{2G\overline{m}}\right]^3. \tag{1.19}$$

The subscript $J$ has been used because these critical values for the mass and density are often called the *Jeans* mass and density.

In fact, it is most useful to express the condition for condensation in terms of the average density of the cloud. We note that the critical density given by Eq. (1.19) is low and hence more easily achieved if the mass of the cloud is large. For example, a cloud of molecular hydrogen at a temperature of 20 K with a mass of $2 \times 10^{33}$ kg, which is equivalent to 1000 solar masses, could condense if its density reaches $10^{-22}$ kg m$^{-3}$; i.e. about $10^5$ molecules per cubic metre. The critical density for a similar cloud with a mass equal to 1 solar mass is a million times higher !

These considerations suggest that it is natural to regard the condensation of gas clouds into stars as taking place in several stages. First, a massive extended gas cloud contracts; its mass may be thousands of times the solar mass. When the cloud has compressed and its density has become high enough, smaller parts of it will be able to contract independently. Ultimately, the cloud will be able to fragment into many parts, each with a mass comparable to the solar mass. These fragments may then condense to form a cluster of primitive stars, *protostars*.

## Contraction of a protostar

Equation (1.19) implies that, when a cloud at a temperature of 20 K reaches a density of $10^{-16}$ kg m$^{-3}$, a fragment with a mass comparable with the solar mass (i.e. $2 \times 10^{30}$ kg) is capable of contracting independently. At this stage the fragment forms a protostar with a radius of the order of $10^{15}$ m, about a million times larger than the sun. It collapses freely, unopposed by internal pressure, if the gravitational energy released is not converted into random thermal motion. This is possible as long as a substantial fraction of the energy released is absorbed by the dissociation of hydrogen molecules and by the ionization of hydrogen atoms.

The energy needed to dissociate a hydrogen molecule is $\epsilon_D = 4.5$ eV, and the energy needed to ionize a hydrogen atom is $\epsilon_I = 13.6$ eV. Hence the energy needed to dissociate and ionize all the hydrogen in a protostar of mass $M$ is approximately

$$\frac{M}{2m_{\mathrm{H}}}\,\epsilon_D + \frac{M}{m_{\mathrm{H}}}\,\epsilon_I,$$

where $m_{\mathrm{H}}$ is the mass of the hydrogen atom. If we assume that this energy is supplied by the gravitational collapse of a protostar from an initial radius $R_1$ to a final radius $R_2$, then

$$\frac{GM^2}{R_2} - \frac{GM^2}{R_1} \approx \frac{M}{2m_{\mathrm{H}}}\,\epsilon_D + \frac{M}{m_{\mathrm{H}}}\,\epsilon_I. \tag{1.20}$$

In particular, the energy needed to dissociate and ionize the hydrogen in a protostar with a mass equal to the solar mass is $3 \times 10^{39}$ J. Such a protostar will collapse freely from its initial radius of $R_1 \approx 10^{15}$ m to a radius $R_2 \approx 10^{11}$ m ; i.e. the radius shrinks 10,000 fold to a size equal to a hundred times the solar radius. The time scale for this collapse is set by Eq. (1.4), which gives the free-fall time for an object of initial density $\rho$. In this case $\rho \approx 10^{-16}$ kg m$^{-3}$ and the time scale is of the order of 20,000 years.

When most of the hydrogen is ionized, and as the protostar becomes increasingly opaque to its own radiation, the gravitational energy released is converted into random thermal energy of electrons and ions. The internal pressure rises and the collapse of the protostar is slowed down, and hydrostatic equilibrium is approached.

It is easy to estimate the average internal temperature of a protostar at the time when the rapid collapse under gravity is replaced by slow contraction. To do so, we use the virial theorem (1.11) to relate the internal kinetic energy and gravitational energy of a protostar when it is near to hydrostatic equilibrium. The thermal kinetic energy of the hydrogen ions and electrons in the protostar at an internal temperature $T$ is

$$E_{KE} \approx \frac{M}{m_{\rm H}} 3kT. \tag{1.21}$$

The gravitational energy at the end of the period of rapid collapse is given by Eq. (1.20); because $R_1 >> R_2$ we have

$$E_{GR} \approx -\frac{GM^2}{R_2} \approx -\left[\frac{M}{2m_{\rm H}}\epsilon_D + \frac{M}{m_{\rm H}}\epsilon_I\right]. \tag{1.22}$$

According to the virial theorem (1.11),

$$2E_{KE} + E_{GR} = 0.$$

Hence, a protostar approaches hydrostatic equilibrium at a temperature given by

$$kT \approx \frac{1}{12}[\epsilon_D + 2\epsilon_I] \approx 2.6 \text{ eV}. \tag{1.23}$$

This corresponds to an average internal temperature of 30,000 K. Note that this estimate is independent of the mass of the protostar.

The subsequent slow contraction of the protostar is governed by the opacity of the ionized interior. This opacity controls the rate at which energy is lost as radiation from the surface, and hence the rate of release of gravitational energy. The time scale for the contraction is of the order of $10^7$ to $10^8$ years. The virial theorem can again be used, because the protostar remains close to a state of hydrostatic equilibrium. According to Eq. (1.11) and Eq. (1.12), half the gravitational energy released is lost from the surface; the other half is stored as internal kinetic energy. The temperature and pressure at the centre of the protostar increase until the conditions are suitable for the thermonuclear fusion of hydrogen. The energy released by nuclear fusion lessens the need for the release of gravitational energy, and the protostar ceases to contract. True stardom is reached when the nuclear reaction rate is sufficient to supply the radiant energy lost from the surface.

### Conditions for stardom

Not all self-gravitating bodies achieve stardom. A hot gas of classical electrons and ions is not the only way to resist gravity. Gravitational contraction can also be opposed by a cold, dense gas of degenerate electrons. In such a gas, the electrons are governed by the laws of quantum mechanics and occupy the lowest possible energy states in accordance with the Pauli exclusion principle. A degenerate electron gas resists compression, not because of random thermal energy of the electrons, but because the total kinetic energy of the electrons has a minimum value which increases as the density rises. In fact, the temperature of a contracting body ceases to rise if the electrons become degenerate. This occurs if the average distance between electrons in the contracting system becomes comparable with the typical de Broglie wavelength of the electrons.

The quantum mechanical de Broglie wavelength of an electron is given by $\lambda = h/p$, where $h$ is Planck's constant and $p$ the momentum. Since the kinetic energy of an electron in a classical gas at temperature $T$ is approximately $kT$, the momentum is about $(m_e kT)^{1/2}$, and the typical de Broglie wavelength is

$$\lambda \approx \frac{h}{(m_e kT)^{1/2}}. \tag{1.24}$$

Classical mechanics will be valid provided the wave functions of the electrons do not overlap; in other words, the average separation between the electrons has to be large compared with $\lambda$. This condition is satisfied if the density of the ionized gas satisfies the inequality

$$\rho << \frac{\overline{m}}{\lambda^3} \approx \overline{m}\,\frac{(m_e kT)^{3/2}}{h^3}. \tag{1.25}$$

Here $\overline{m}$ is the average mass of the particles in the ionized gas; for ionized hydrogen $\overline{m} = 0.5$ amu, the average mass of a proton and an electron.

It is straightforward to show that the internal temperature of the protostar will initially rise as the internal density increases. Substitution of the approximate expression for the gravitational energy, Eq. (1.16), and the classical expression for the internal kinetic energy, Eq. (1.17), into the condition for hydrostatic equilibrium, $2E_{KE} + E_{GR} = 0$, gives

$$kT \approx \frac{GM\overline{m}}{3R} \approx G\overline{m}M^{2/3}\rho^{1/3}. \tag{1.26}$$

We see that the temperature is proportional to $\rho^{1/3}$. This will be the case as long as the density is low enough to satisfy the inequality (1.25) so that the electrons are governed by classical mechanics.

When the density reaches the value

$$\rho \approx \overline{m} \, \frac{(m_e kT)^{3/2}}{h^3}, \tag{1.27}$$

quantum mechanics becomes important and the electrons will begin to become degenerate. As a result the temperature of the gas no longer increases markedly if it is compressed. We can estimate the temperature at which the electrons in the contracting protostar become degenerate by substituting the critical density given by Eq. (1.27) into Eq. (1.26). We obtain

$$kT \approx G\overline{m}M^{2/3}\overline{m}^{1/3}\frac{(m_e kT)^{1/2}}{h},$$

which can be rearranged to give

$$kT \approx \left[\frac{G^2\overline{m}^{8/3}m_e}{h^2}\right] M^{4/3}. \tag{1.28}$$

Around this temperature degenerate electrons begin to resist compression and further contraction under gravity no longer causes the temperature to rise.

Equation (1.28) gives an estimate for the maximum value of the average internal temperature reached by a contracting protostar. Notice the key role played by the mass $M$ of the protostar. If the solar mass of $2 \times 10^{30}$ kg is substituted we obtain a maximum temperature of $kT \approx 1$ keV. In other words, a solar mass, if it were allowed to contract under gravity, could reach an average internal temperature of about 10 million K, and a central temperature which is even higher; this is more than sufficient to trigger thermonuclear reactions and the fusion of hydrogen to helium. But the contraction of protostars less massive than the sun lead to lower internal temperatures. Detailed calculations indicate that the minimum mass needed for thermonuclear ignition, and hence true stardom, is about 0.08 solar masses. Protostars with masses less than this value evolve into objects where gravity is countered by the pressure of degenerate electrons; such objects are often called brown dwarfs.

We shall consider the possible range of masses for stars in Chapter 5. The minimum mass for a star will be examined in more detail, and we shall also argue that there is a maximum as well as a minimum mass for stardom. In particular, it will be shown that the pressure generated by radiation inside a star is significant if the mass is much larger than the solar mass. This implies that the hydrostatic equilibrium of a massive star is dependent on radiation pressure, i.e.

on the pressure due to a gas of photons. But, as we have already seen in Section 1.2, hydrostatic equilibrium becomes precarious as the gas particles become ultra-relativistic: according to Eq. (1.13) the binding energy becomes small, and small changes in the total energy are accompanied by large changes in the internal and gravitational energies. These considerations suggest that stars with a mass greater than 50 to 100 solar masses are easily disrupted. And indeed such stars are rare.

## 1.4 THE SUN

As our nearest star the sun has a special role as a source of precise astrophysical information. For example, we know its mass, radius, geometric shape and age, and also the luminosity and spectrum of electromagnetic radiation from its surface. This observational information is used in theoretical models of the sun to predict the physical characteristics of the solar interior. The most detailed model of the sun is the Standard Solar Model which is described by Bahcall (1989). Some of the input parameters for this model and some of the calculated solar properties are listed in Table 1.2.

Our aim in this section is to consider the sun in its simplest terms in order to illustrate basic astrophysical concepts and to fix the order of magnitude of astrophysical quantities.

TABLE 1.2 The main physical properties of the sun. The measured properties are the mass, radius, oblateness, photon luminosity, and surface temperature. The estimate for the age is largely based on geological studies. The properties at the centre of the sun are calculated with the aid of the Standard Solar Model; see Bahcall (1989) for more detail.

| Property | Value |
|---|---|
| Mass | $M_\odot = 1.99 \times 10^{30}$ kg |
| Radius | $R_\odot = 6.96 \times 10^{8}$ m |
| Photon luminosity | $L_\odot = 3.86 \times 10^{26}$ W |
| Effective surface temperature | $T_E = 5780$ K |
| Age | $t_\odot \approx 4.55 \times 10^{9}$ years |
| Central density | $\rho_c = 1.48 \times 10^{5}$ kg m$^{-3}$ |
| Central temperature | $T_c = 15.6 \times 10^{6}$ K |
| Central pressure | $P_c = 2.29 \times 10^{16}$ Pa |

### Pressure, density and temperature

The sun is a star of mass $M_\odot \approx 2 \times 10^{30}$ kg. The gravitational contraction of the sun was halted about 5 billion years ago by the ignition of 'hydrogen burning'; i.e.

the thermonuclear fusion of hydrogen to form helium. During its current hydrogen burning phase the solar radius is $R_\odot \approx 7 \times 10^8$ m and the average density $\langle\rho\rangle$ is $1.4 \times 10^3$ kg m$^{-3}$. The time for free fall under gravity for an object of this density is given by Eq. (1.4),

$$t_{FF} = \left[\frac{3\pi}{32G\langle\rho\rangle}\right]^{1/2} \approx \frac{1}{2} \text{ hour.}$$

As this time bears no relation to the sun we observe, we safely conclude that the sun is not in free fall and that the internal pressure gradient within the sun must play an essential role in opposing gravity. Indeed, as there is no evidence for major changes in the sun during the geological lifetime of the earth, we can conclude that the sun has been close to hydrostatic equilibrium for at least 4.5 billion years. Hydrostatic equilibrium implies we can use the virial theorem to find the average pressure supporting the sun; using Eqs. (1.7) and (1.16) we find

$$\langle P\rangle = -\frac{1}{3}\frac{E_{GR}}{V} \approx \frac{GM_\odot^2}{4\pi R_\odot^4} \approx 10^{14} \text{ Pa.} \tag{1.29}$$

Hence the interior of the sun provides an environment in which matter and radiation interact at high temperature such that, on average, the pressure is about a billion times atmospheric pressure and the density is comparable with normal water. The thermal physics needed to understand matter and radiation under these extreme conditions will be reviewed in Chapter 2; the ionization of gases and the equations of state for non-relativistic, ultra-relativistic, classical and quantum gases will be discussed. This discussion indicates that we are justified in making the simple and bold assumption that the sun is primarily supported by the pressure of an ideal classical gas of electrons and ions. Thus, the average pressure inside the sun is given by

$$\langle P\rangle = \frac{\langle\rho\rangle}{\overline{m}}kT_I, \tag{1.30}$$

where $T_I$ is the typical internal temperature and $\overline{m}$ is the average mass of the gas particles. For ionized hydrogen $\overline{m} = 0.5$ amu, the average mass of a proton and an electron. In fact, the Standard Solar Model assumes that the sun was formed from material which was 71% hydrogen, 27% helium and 2% of heavy elements, such as carbon, oxygen and iron. When fully ionized this yields an average gas particle mass of $\overline{m} \approx 0.61$ amu.

It is easy to combine Eqs. (1.29) and (1.30) to estimate the typical temperature inside the sun. We obtain

$$kT_I \approx \frac{GM_\odot\overline{m}}{3R_\odot} \approx 0.5 \text{ keV} \quad \text{or} \quad T_I \approx 6 \times 10^6 \text{ K.} \tag{1.31}$$

Of course the actual temperature inside the sun, like the density and the pressure, increases towards the centre. The central temperature, density and pressure given by the Standard Solar Model are listed in Table 1.2.

### Solar radiation

The total power radiated by the sun, its luminosity $L_\odot$, is about $4 \times 10^{26}$ W. Moreover, to a first approximation the sun appears to be a black body radiator with area $4\pi R_\odot^2$ and effective surface temperature $T_E$ of about 6000 K. Thus

$$L_\odot = 4\pi R_\odot^2 \sigma T_E^4, \tag{1.32}$$

where $\sigma$ is Stefan's constant, $5.67 \times 10^{-8}$ W m$^{-2}$ K$^{-4}$. Since $kT_E \approx 0.5$ eV the bulk of the radiation is in the visible part of the electromagnetic spectrum.

We note that the effective surface temperature, $T_E \approx 6000$ K, is three orders of magnitude less than the typical interior temperature of $T_I \approx 6{,}000{,}000$ K given by Eq. (1.31). We can understand this difference by examining the mechanism by which radiation escapes from the sun.

As the electrons and ions interact inside the sun they emit electromagnetic radiation which in turn interacts with electrons and ions. Indeed, to a first approximation we can consider the sun as a globe of electrons and ions in equilibrium with electromagnetic radiation at a temperature $T_I$. If this radiation were free to escape, without disturbing thermodynamic equilibrium, then the sun would appear to be a black body radiator at a temperature of $T_I$. The luminosity of the sun would be

$$L'_\odot \approx 4\pi R_\odot^2 \sigma T_I^4, \tag{1.33}$$

and the radiation would be in the X-ray region of the electromagnetic spectrum because $kT_I \approx 0.5$ keV.

Fortunately for the inhabitants of planet earth, this radiation is not free to escape; to a very large extent it is trapped within an opaque sun and the earth is not incinerated by X-rays. The radiation inside the sun is continually scattered, absorbed and emitted by electrons and ions. A temperature gradient is set up and the radiant energy slowly diffuses towards the surface where it escapes as visible radiation. The underlying mechanism for radiative diffusion is a random walk in which the photons are scattered, absorbed and emitted, as shown in Fig. 1.2.

We shall let $l$ represent a free path for a photon within the sun. In practice there is a distribution of free paths with a mean which depends on the region within the sun. To keep the analysis as simple as possible we will take $l$ to be a constant length characteristic of photons within the sun as a whole. After $N$ interactions, and after $N$ vector displacements in random directions, the radiant energy associated with the photon has travelled a vector distance

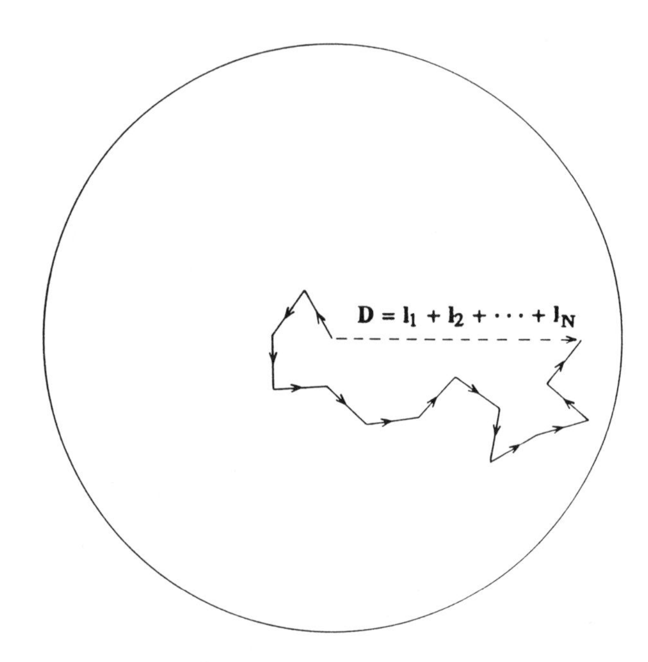


Fig. 1.2 A random walk mechanism for radiative diffusion. A sequence of $N$ steps in random directions leads to a vector displacement of $\mathbf{D} = \mathbf{l}_1 + \mathbf{l}_2 + \cdots + \mathbf{l}_N$.

$$\mathbf{D} = \mathbf{l}_1 + \mathbf{l}_2 + \cdots + \mathbf{l}_N,$$

as shown in Fig. 1.2. The square of the net distance travelled in $N$ steps is

$$D^2 = l_1^2 + l_2^2 + \cdots + l_N^2 + 2(\mathbf{l}_1 \cdot \mathbf{l}_2 + \mathbf{l}_1 \cdot \mathbf{l}_3 + \cdots)$$

If we average over many random walks, the terms involving scalar products cancel because the direction of each step is random. Hence the mean square distance travelled is simply $l_1^2 + l_2^2 + \cdots + l_N^2$ or $Nl^2$.

To escape from the sun, a photon must diffuse a distance which is comparable with the solar radius. On average this requires about $R_\odot^2/l^2$ steps. Because the time for each step is $l/c$, where $c$ is the speed of light, the random-walk escape time is approximately

$$t_{RW} \approx \frac{R_\odot^2}{cl}. \tag{1.34}$$

In contrast, the time to escape directly from the sun is $R_\odot/c$, which is a factor of $l/R_\odot$ shorter than the random-walk escape time. Thus, radiative diffusion via a random walk slows down the rate at which energy escapes from the sun by a factor of $l/R_\odot$. It follows that the actual luminosity of the sun given by Eq. (1.32) is a factor of $l/R_\odot$ smaller than the luminosity given by Eq. (1.33), the luminosity that would arise if radiation were free to escape unhindered. This implies that the effective surface temperature and the typical internal temperature of the sun are approximately related by

$$T_E \approx \left[\frac{l}{R_\odot}\right]^{1/4} T_I. \tag{1.35}$$

Using $T_E \approx 6000$ K and $T_I \approx 6{,}000{,}000$ K, we find that the effective mean free path for radiative diffusion in the sun is about 1 mm; i.e. the sun is very opaque. And using Eq. (1.34), we find that the typical time for radiation to diffuse from the centre and escape from the sun is about 50,000 years.

We can also use this simple but approximate analysis to reveal how the luminosity of a star like the sun depends on its mass. An approximate expression for the luminosity can be found (Eq. 1.32) by using Eq. (1.35) and the relation between the internal temperature and the mass and radius of the sun given in Eq. (1.31). We find

$$L_\odot \approx 4\pi R_\odot^2 \sigma T_I^4 \frac{l}{R_\odot} \approx \frac{(4\pi)^2}{3^5}\frac{\sigma}{k^4} G^4\, \overline{m}^4 \langle\rho\rangle l\, M_\odot^3. \tag{1.36}$$

This equation indicates that the luminosity of a sun-like star is expected to be a rapidly increasing function of its mass.

Radiative diffusion will be considered in more detail in Chapter 3. We shall end this preliminary discussion by emphasizing that radiative diffusion restricts the flow of radiation and prevents the sun from losing heat catastrophically. It determines the luminosity and hence the rate at which energy must be released by thermonuclear fusion at the centre of the sun.

### Thermonuclear fusion in the sun

Thermonuclear fusion will be considered in detail in Chapter 4. At this point we note that the solar luminosity is currently being supplied by a chain of thermonuclear reactions called the proton–proton chain. The dominant reactions are:

$$p + p \rightarrow d + e^+ + \nu_e, \tag{1.37}$$

$$p + d \rightarrow {}^3\mathrm{He} + \gamma, \tag{1.38}$$

$${}^3\mathrm{He} + {}^3\mathrm{He} \rightarrow {}^4\mathrm{He} + p + p, \tag{1.39}$$

where $d$ denotes a deuteron or $^2$H, an isotope of hydrogen with mass 2. Each of these reactions is exothermic and the total thermonuclear energy release is about 26 MeV per $^4$He nucleus formed. This energy must be released at a rate of $4 \times 10^{26}$ W in order to power the solar luminosity.

All the reactions in the proton–proton chain are hindered because a Coulomb barrier tends to keep the positively charged nuclei apart. However, there is a significant probability that nuclei can tunnel, quantum mechanically, through a Coulomb barrier if the temperature is high. The interaction required to effect fusion is different for each of the reactions; Eq. (1.37) relies on the weak nuclear interaction, Eq. (1.38) relies on the electromagnetic interaction and Eq. (1.39) relies on the strong nuclear interaction. As a result, the first reaction in the chain, Eq. (1.37), is by far the slowest. As we shall see in Chapter 4, a proton at the centre of the sun takes, on average, about 5 billion years before it fuses with another proton to produce a deuteron. The deuteron so produced is snapped up to form a $^3$He in about a second and the average time needed for two $^3$He to collide and form a $^4$He nucleus is approximately 300,000 years.

It follows that the first reaction in the chain, the slow weak reaction Eq. (1.37), governs the rate at which energy is released by the proton–proton chain. This reaction forms a bottle-neck through which an immense store of hydrogen fuel is gradually processed. One consequence is that, even though the total power released is huge, $4 \times 10^{26}$ W, the power density is very modest. On average each kilogram of the sun generates only 0.2 of a milliwatt; this is about 10,000 times less than the power density generated by the metabolic activity in the human body.

We note that the weak reaction, Eq. (1.37), implies that, as protons are consumed, neutrinos are emitted. Four protons are needed to produce a $^4$He nucleus and release 26 MeV, i.e. $26 \times 1.6 \times 10^{-13}$ J. Hence the rate of consumption of protons needed to power a solar photon luminosity of $4 \times 10^{26}$ W is

$$(4 \times 4 \times 10^{26})/(26 \times 1.6 \times 10^{-13}), \quad \text{or} \quad 4 \times 10^{38} \text{ protons per second.}$$

The fusion of these protons is also accompanied by the emission of at least $2 \times 10^{38}$ neutrinos per second. These weakly produced neutrinos can subsequently interact but only weakly via the weak nuclear interaction. Unlike photons, they pass through the sun and escape almost unhindered and, if detected on earth, they could provide direct inside information on the thermonuclear reactions occurring at the centre of the sun. Needless to say, the detection on earth of particles which can pass almost unhindered through the sun is a formidable exercise. The detection of solar neutrinos and the solar neutrino problem will be considered in Chapter 4.

Thermonuclear fusion not only postpones the contraction of the sun, it also acts as a solar temperature regulator, a thermonuclear thermostat. If the temperature rises, the nuclear reaction rate will increase and release more energy than can escape. Because the sun remains close to hydrostatic equilibrium we can apply the virial theorem to see what happens when the total energy is increased in this way.

Equations (1.11) and (1.12) show that there will be an increase in the gravitational energy and a decrease in the internal energy; in other words the sun will expand and cool. A parallel set of events will occur if the temperature falls: the energy released by the nuclear reactions will not be high enough to supply the energy lost by the sun and the total energy will be reduced, and this reduction in the total energy will cause the sun to contract and heat up.

This thermonuclear thermostat has postponed gravitational contraction and kept the sun steady for at least 4.5 billion years. It will continue to do so until there is insufficient hydrogen at the centre of the sun to fuel the proton–proton chain and supply the required solar luminosity of $4 \times 10^{26}$ W. There are approximately $7 \times 10^{56}$ protons in the sun, and, as they are being consumed at a rate of $4 \times 10^{38}$ every second, 10% will be consumed in the next 6 billion years. In total, the hydrogen burning phase of the sun will last for about 10 billion years, after which the central core of the sun will contract and heat up until the temperature and density are high enough to ignite the thermonuclear fusion of helium. The outer layers of the sun will expand to form a red giant, and the sun will begin its next stage of stellar evolution.

## 1.5 STELLAR NUCLEOSYNTHESIS

Stellar evolution involves the release of gravitational potential energy through contraction, with pauses whenever nuclear fuels are ignited so as to supply the energy flow from the surface of the star. The ashes of one set of nuclear reactions may become the fuel for the next set. For example, the helium produced by the fusion of hydrogen may be ignited in a subsequent gravitational contraction to produce carbon. In fact, there is a sequence of thermonuclear stages. Each stage can be effective in calling a temporary halt to gravitational contraction provided it leads to the release of energy through the formation of more tightly bound nuclei.

The binding energy per nucleon for atomic nuclei is illustrated in Fig. 1.3. The broad maximum at a mass number near 56 implies that the nuclei near iron in the periodic table are the most tightly bound. Thus, the sequence of thermonuclear reactions in stars is expected to terminate when nuclei near iron are produced. These nuclei, isotopes of Cr, Mn, Fe, Co and Ni, form a nuclear ash which cannot be burnt.

The main stages of thermonuclear fusion in stars and the approximate temperature needed to ignite each stage are listed in Table 1.3.

### Stellar mass and the extent of thermonuclear fusion

Not all stars can achieve the temperatures needed to ignite every stage of thermonuclear fusion and progress to the synthesis of iron. We recall that the internal temperature of a contracting star ceases to rise when the electrons within the star become degenerate; i.e. when the average distance between the electrons

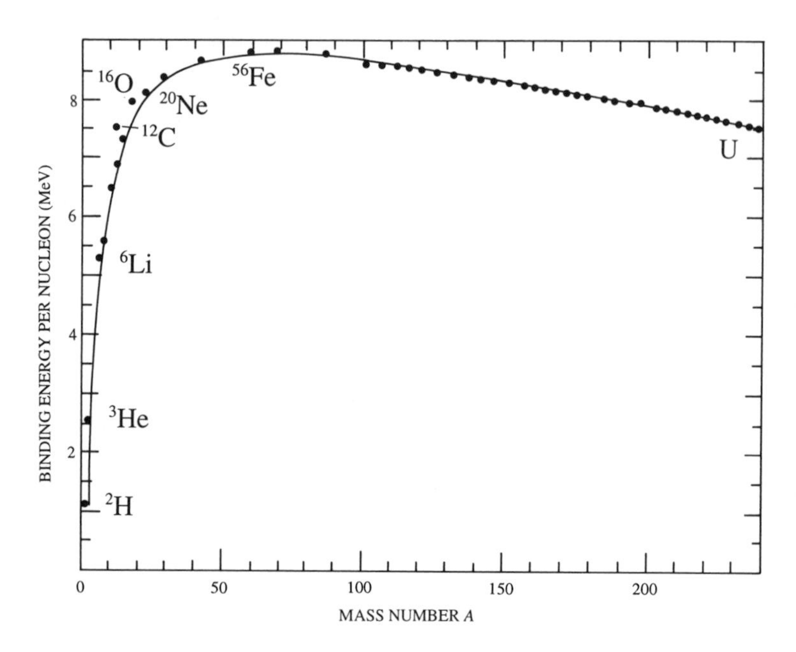


Fig. 1.3 Binding energy per nucleon for atomic nuclei. There is a broad maximum at mass number 56 which implies that energy is normally released when two light nuclei fuse to form a heavier nucleus provided the nucleus formed has a mass number less than 56.

TABLE 1.3 The main stages of nuclear burning in stars. The ashes of one stage of burning may become the fuel for the next stage provided the contracting star is massive enough to reach the approximate ignition temperature indicated.

| Process | Fuel | Products | Approximate ignition temperature |
|---|---|---|---|
| Hydrogen burning | Hydrogen | Helium | $1 \times 10^7$ K |
| Helium burning | Helium | Carbon<br>Oxygen | $1 \times 10^8$ K |
| Carbon burning | Carbon | Oxygen<br>Neon<br>Sodium<br>Magnesium | $5 \times 10^8$ K |
| Neon burning | Neon | Oxygen<br>Magnesium | $1 \times 10^9$ K |
| Oxygen burning | Oxygen | Magnesium<br>to sulphur | $2 \times 10^9$ K |
| Silicon burning | Silicon | Iron and<br>nearby elements | $3 \times 10^9$ K |

becomes comparable with the typical de Broglie wavelength of the electrons. In particular we found in Section 1.3 that the maximum temperature possible is approximately proportional to $M^{4/3}$; see Eq. (1.28). Thus the mass of a contracting star determines the maximum temperature achievable and hence which thermonuclear fusion stages can be reached.

We have already mentioned that only stars with a mass greater than $0.08M_\odot$ can attain true stardom and ignite hydrogen. There are in fact two mechanisms for hydrogen burning. The proton–proton chain is important in stars like the sun. But in more massive stars hydrogen is fused to helium via a set of reactions in which carbon acts as a catalyst, the so-called carbon–nitrogen cycle; an important by-product of this type of hydrogen burning is nitrogen. These, and other aspects of thermonuclear fusion, will be considered in more detail in Chapter 4.

When hydrogen burning ceases in the centre of the star, the helium core contracts under gravity and grows hotter. The increased temperature promotes hydrogen burning in a shell surrounding the core. It also leads to an increase in pressure and a large expansion of the outer layers of the star. As hydrogen burning continues in the shell, more helium is produced and deposited onto a helium core which becomes hotter and denser. If the star is massive enough, the core becomes sufficiently hot and dense for helium nuclei to fuse together to form carbon nuclei. Helium burning releases energy which causes the core to expand and cool, and a cooler core leads to a partial contraction of the outer layers of the star. The star is now a red giant with a luminosity dominantly powered by helium burning in a hot, dense central core. The temperature and density of this core is between 1 and $2 \times 10^8$ K and $10^5$ and $10^8$ kg m$^{-3}$. To achieve these conditions the initial mass of the star must exceed a value of about $0.5M_\odot$.

In fact, helium burning is severely hindered by the absence of stable nuclei with mass 5 and mass 8. The only way forward involves the fusion of three $^4$He nuclei to produce $^{12}$C via the three-body reaction:

$$^4\text{He} + {}^4\text{He} + {}^4\text{He} \rightarrow {}^{12}\text{C}.$$

This three-body reaction actually takes place in two stages: Two $^4$He nuclei fuse to form an unstable $^8$Be nucleus whose brief existence is just long enough to permit an occasional capture of a third $^4$He nucleus to form $^{12}$C. This requires both a high density and a high temperature. We note that helium was produced but not burnt in big bang nucleosynthesis; the temperature was hot enough but the density was too low. The high density and temperature needed for helium burning had to await the formation and evolution of massive stars.

Helium burning not only produces carbon it also leads to another vitally important element, namely oxygen, via the reaction

$$^4\text{He} + {}^{12}\text{C} \rightarrow {}^{16}\text{O} + \gamma.$$

In addition, small amounts of $^{20}$Ne are also formed by

$$^{4}\mathrm{He} + ^{16}\mathrm{O} \rightarrow ^{20}\mathrm{Ne} + \gamma.$$

As helium is consumed in the centre of the star, helium burning migrates to a shell surrounding a central core of carbon and oxygen, leading to an onion-like structure for the star in which there is an outer hydrogen burning layer, an inner helium burning layer and a core of carbon and oxygen; the outer layers of the star expand markedly during this phase of evolution.

Stars with a mass greater than $8M_\odot$, or thereabouts, can progress beyond helium burning and ignite carbon at a temperature of about $5 \times 10^8$ K to form elements such as neon, sodium and magnesium. If the temperature exceeds $10^9$ K, carbon burning can be followed by the photodisintegration of neon to produce oxygen and helium nuclei; the helium nuclei are then captured by undissociated neon nuclei to form magnesium. Oxygen burning can then take place at about $2 \times 10^9$ K to produce elements between magnesium and sulphur. Stars with a mass greater than $11M_\odot$, or thereabouts, are able to achieve the high temperature of about $3 \times 10^9$ K which is necessary for the ignition of the final stage of thermonuclear fusion. This is silicon burning which leads to the formation of nuclei near iron in the periodic table. Such stars develop a structure consisting of concentric layers composed mostly of hydrogen, helium, carbon, neon, oxygen and silicon surrounding a core of iron and nearby elements.

In summary, the mass of a star governs the extent to which it converts hydrogen to heavier elements. Contracting stars with a mass approximately between $0.1M_\odot$ and $0.5M_\odot$ will reach the required temperature to ignite hydrogen but they do not get hot enough to ignite helium. Stars with mass roughly between $0.5M_\odot$ and $8M_\odot$ will ignite hydrogen and helium, and stars in the mass range around 8 to $11M_\odot$ will progress beyond helium burning to carbon burning. Finally, stars with a mass greater than $11M_\odot$ are able to achieve the high temperatures necessary for the ignition of every stage of thermonuclear fusion.

**Neutron capture**

Thermonuclear fusion provides a mechanism for the release of energy and the production of elements up to iron in the periodic table. We also need a mechanism to account for the existence of elements heavier than iron. In general, energy is needed to produce these elements and fusion of charged nuclei is not effective. These elements owe their existence to neutron capture.

Neutrons are released by nuclear collisions and photodisintegration, particularly during the later stages of stellar evolution. Because neutrons are electrically neutral, they are easily captured by a nucleus to form a more massive nucleus with the same charge. Thus the presence of neutrons can lead to the production of neutron-rich isotopes. Such isotopes will eventually decay by beta decay; a neutron within the nucleus is converted into a proton and the atomic number of the nucleus increases by one unit. It is believed that the elements heavier than iron have been produced by sequences of neutron capture reactions followed by sequences of beta decays.

The production of neutrons in an evolved star is normally a slow process, and any nucleus formed by neutron capture will have plenty of time to beta decay. This process of forming atomic nuclei is called the *s-process* where *s* stands for slow. However, neutron production may become very rapid during the final stage of evolution of a massive star. We shall see that this stage involves the collapse of a central core of iron which, amongst other things, can lead to the ejection of the outer layers of the star to form a supernova. During this explosive stage, nuclei can capture many neutrons before beta decay becomes effective. This process is called the *r-process* where *r* stands for rapid. The types of nuclei produced by these two processes differ significantly. For example, no element beyond bismuth ($Z = 83$) can be formed by the s-process, whereas the r-process can produce elements beyond this.

## 1.6 STELLAR LIFE CYCLES

The big bang led to a universe composed of hydrogen and helium with traces of light elements. This primordial matter has been enriched with heavier elements by a cycle of stellar formation and evolution in which matter has been transferred back and forth between stars and interstellar matter. One of the main aims of astrophysics is to use this cycle to explain the abundances of the chemical elements in the universe today.

### Rate of stellar evolution

In our discussion of the sun in Section 1.4 we saw that the luminosity of the sun determines the rate at which it consumes its nuclear fuel. In particular, Eq. (1.36) indicates that the luminosity of a star is a rapidly increasing function of its mass. Indeed, if the mean free path for radiative diffusion, $l$, is inversely proportional to the density, the luminosity given by Eq. (1.36) is proportional to the cube of the mass of the star. Figure 1.4 illustrates the actual relation between the mass $M$ and luminosity $L$ of representative hydrogen burning stars like the sun. We note that the luminosity is proportional to $M^{\alpha}$, where $\alpha$ is about 3 for massive stars and about 3.5 for stars less massive than the sun.

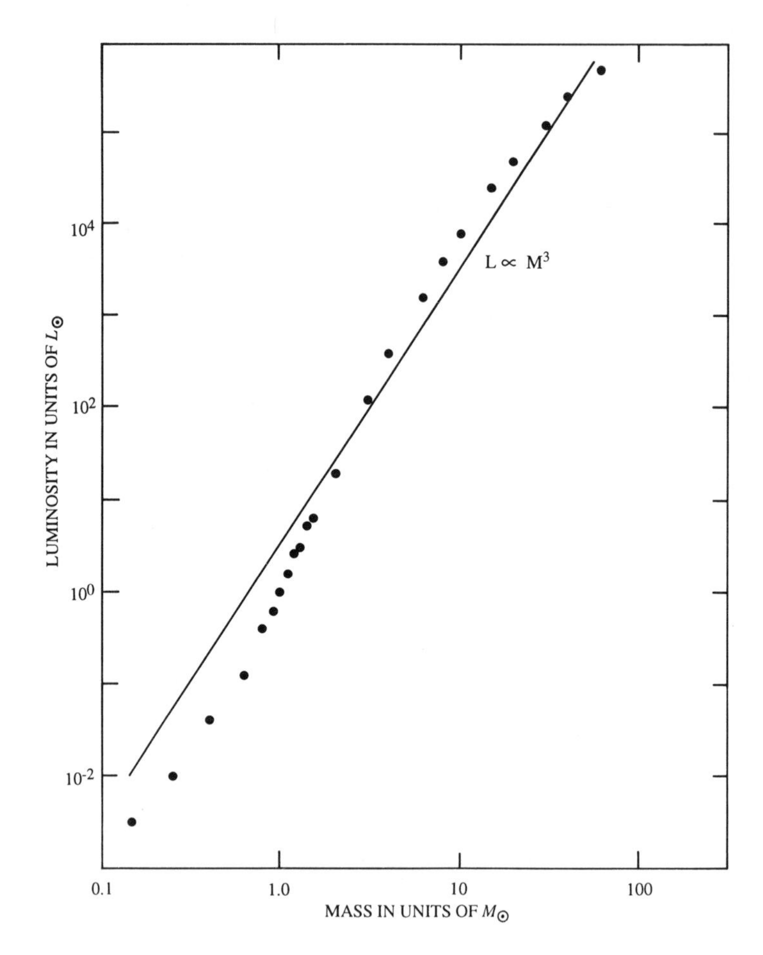


Fig. 1.4 The mass–luminosity relation for hydrogen burning stars with a chemical composition similar to the sun. The data on representative main sequence stars is taken from Table 3.13 in the Astronomy and Astrophysics section of the *Physics Vade Mecum* compiled by Fredrick (1989).

This rapid increase of luminosity with mass has an important implication: it implies that massive stars have shorter lives despite their greater resource of fuel. Since the fuel reserves are proportional to $M$, the hydrogen burning lifetime is proportional to $M^{-2}$ for high mass stars and $M^{-2.5}$ for low mass stars. Given that the hydrogen burning life of the sun is about 10 billion years, we conclude that a star of mass $10M_\odot$ will burn hydrogen for about 100 million years, whereas the hydrogen burning lifetime of a star of mass $0.5M_\odot$ will exceed 50 billion years.

In fact, the overall rate for all nuclear fusion processes inside a star, and hence the rate at which the star evolves, is largely determined by its mass; more massive stars evolve more quickly. Since the lifetime of the universe is between 10 to 20 billion years, there has been ample time for many generations of massive stars, but there has been insufficient time for the evolution of stars with a mass much smaller than the sun.

### The end-points of stellar evolution

The ultimate fate of a star depends crucially on the mass that remains in the central core when nuclear fusion can no longer maintain the pressure needed to prevent gravitational contraction. At this stage, the star must rely on a non-thermal source of pressure for support, namely a gas of degenerate electrons. However, we recall that hydrostatic equilibrium becomes precarious if gravity is opposed by the pressure generated by ultra-relativistic particles. This general principle imposes an upper limit on the mass that can be supported by a degenerate electron gas. In particular, we shall show in Chapter 6 that if the mass of a stellar core exceeds a critical value, the degenerate electrons become sufficiently relativistic to render hydrostatic equilibrium impossible. This critical mass is about $1.4M_{\odot}$ and is called the Chandrasekhar mass.

Thus the fate of an evolved star depends crucially on whether the mass of its central core is less than or greater than the Chandrasekhar mass. A star like the sun will develop a stellar core with a mass less than $1.4M_{\odot}$, which can be supported by the pressure of degenerate electrons. After it loses its outer tenuous layers it forms a white dwarf, a compact object with a radius of about $10^7$ m and a density of about $10^9$ kg m$^{-3}$, which slowly cools without appreciable contraction because its mechanical support is due to a pressure which is insensitive to temperature.

As already mentioned, massive stars develop an onion-like structure with a central core of iron. The mass of this inert core grows as silicon burning deposits more iron. Eventually the core will collapse catastrophically when its mass exceeds the Chandrasekhar limit; this collapse is considered in Section 6.2 of Chapter 6. To a first approximation the collapse is a free fall under gravity, unopposed by an internal pressure gradient because energy is absorbed by processes such as the photodisintegration of iron and inverse beta decay. The bulk of the gravitational energy released is carried away by a pulse of neutrinos. But a small fraction of this gravitational energy may be used to eject a substantial fraction of the stellar mass into interstellar space to form a supernova. Stellar nucleosynthesis is completed during these final stages of stellar evolution. In particular, elements heavier than iron are produced by neutron capture.

The eventual mass of the collapsed core is crucial to the final outcome of the evolution of a massive star. The most likely result is the formation of a neutron star, a compact star consisting primarily of degenerate neutrons. There is a maximum possible mass for such an object which is analogous to the Chandrasekhar mass for a star supported by degenerate electrons. This limit is discussed in Section 6.3 of Chapter 6. It is probably about $3M_{\odot}$, but the exact value is uncertain because of the uncertain compressibility of nuclear matter at high densities. It is thought that if the mass of the collapsed core at the heart of a supernova exceeds this limit there is no possibility of halting gravitational collapse. A black hole is produced.

One of the uncertainties in tracing the evolution of a star is the uncertainty in the amount of matter ejected into interstellar space as the star evolves. This mass loss can affect both the rate and the ultimate destination of stellar evolution. Stars lose matter even during the hydrogen-burning phase of evolution; the solar wind, for example, carries away about $10^{-13}M_{\odot}$ of the solar mass every year. As stars evolve, even more intense flows occur as the tenuous outer layers expand. Further, the final stages of evolution are often characterized by significant mass loss. As intermediate mass stars, like the sun, exhaust their nuclear fuel, they shed their outer layers in an expanding cloud called a planetary nebula. In contrast, more massive stars often end their lives with an explosive ejection of matter in a supernova. This matter, together with the matter ejected as planetary nebulae by less massive stars and the matter lost during the earlier stages of stellar evolution, then forms the raw material for future generations of stars.

## Abundances of the chemical elements

The cycle of stellar formation, evolution and death has led to an enrichment of the primordial hydrogen and helium with heavier elements. In particular, the chemical elements observed in the solar system are largely a reflection of the combined effect of nucleosynthesis during the big bang and of nucleosynthesis during the stellar evolution of earlier generations of nearby stars.

The relative abundances of elements in the solar system are plotted against the atomic number $Z$ in Fig. 1.5. The most notable features are:

- The dominance of hydrogen and helium, largely a left-over from nucleosynthesis during the big bang.
- A distinct lack of abundance between helium and carbon, reflecting the difficulty of building elements from hydrogen and helium in the absence of stable mass 5 and mass 8 atomic nuclei.
- Peaks corresponding to the major products of stellar nucleosynthesis; namely carbon, oxygen, neon, silicon and elements near iron. The high abundance of nitrogen, the element between carbon and oxygen, is due to hydrogen burning by the carbon–nitrogen cycle.

In general, thermonuclear fusion, i.e. hydrogen, helium, carbon, oxygen, neon and silicon burning, is responsible for the abundances of elements with atomic number in the range $12 < Z < 30$. Elements with atomic number $Z > 30$ owe their existence to neutron capture particularly during the terminal stages of stellar evolution. In addition, small quantities of elements throughout the periodic table are produced by cosmic ray collisions; indeed a substantial proportion of the elements between helium and carbon have been formed in this way.

For each element there are often several naturally occurring isotopes. Their relative abundance provides further insight into the mechanisms of nucleosynthesis. In addition, some of these isotopes are unstable. Indeed, the continuing presence of radionuclides, such as $^{235}U$ , $^{238}U$ and $^{40}K$ , all with lifetimes comparable with $10^9$ years, enables us to estimate that the solar system was formed some 4.5 billion years ago.

## 1.7 THE HERTZSPRUNG–RUSSELL DIAGRAM

We shall end this introductory chapter by briefly considering some observational properties of stars. It is important to note that stars are opaque to electromagnetic radiation and astronomers are therefore limited to recording superficial information. Moreover, the angular size of even the nearest stars is only a few thousandths of a second. Hence, with rare exceptions, a star appears as a point source of radiation from which the observer can deduce a luminosity and a surface temperature.

### Luminosity

The observed brightness of a star is usually expressed as a *magnitude*. The faintest stars visible to the naked eye have a magnitude of 6, and brighter stars have a smaller magnitude. The scale is logarithmic such that each 10-fold increase in brightness decreases the magnitude by 2.5. Thus if the energy flux received from two stars is $f_1$ and $f_2$, the magnitudes differ by

$$m_1 - m_2 = -2.5 \log_{10}(f_1/f_2). \tag{1.40}$$

Astronomers worry about absolute, apparent, visual and bolometric magnitudes. The absolute bolometric magnitude, $M_{BOL}$, is the most difficult to determine and the most useful. It corresponds to the brightness of a star as measured at a distance of 10 parsecs by a hypothetical detector which responds to the entire electromagnetic spectrum.

The *parsec* is the standard astronomical unit of distance. It is the distance at which one second of arc is subtended by a baseline whose length is equal to the mean separation of the earth and the sun. The numerical value of a parsec is

$$1 \text{ pc} = 3.086 \times 10^{16} \text{ m} = 3.26 \text{ light years}. \tag{1.41}$$

We note that the accurate determination of distance has always been and remains today one of the central problems in astronomy, and that such a determination is needed to deduce the absolute bolometric magnitude of a star.

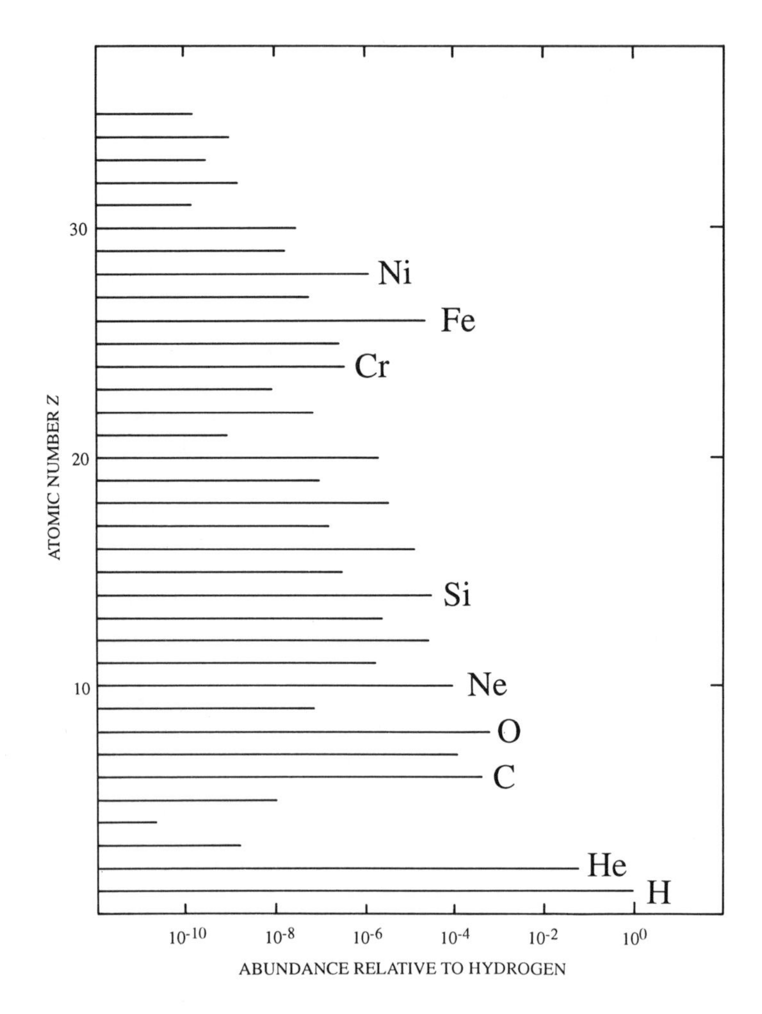


Fig. 1.5 The abundances of elements in the solar system relative to the abundance of hydrogen. Data is taken from Table 3.06 in the Astronomy and Astrophysics section of the *Physics Vade Mecum* compiled by Fredrick (1989). The dominance of hydrogen and helium is a result of nucleosynthesis during the big bang. Thermonuclear fusion in stars preferentially produces helium, carbon, oxygen, neon, silicon and elements near iron. The abundances of elements beyond iron in the periodic table are low; e.g. the abundances of silver, gold and lead relative to hydrogen are $1 \times 10^{-11}$, $6 \times 10^{-12}$ and $1 \times 10^{-10}$. Elements like these are produced in the latter stages of stellar evolution by the capture of neutrons by lighter nuclei followed by beta-decay.

Because the absolute bolometric magnitude $M_{BOL}$ represents the brightness of a star at a specific distance, it provides an absolute measure of the luminosity. In fact, a star with luminosity $L$ has an absolute bolometric magnitude given by

$$M_{BOL} = -2.5 \log_{10}(L/L_\odot) + 4.72, \tag{1.42}$$

where $L_\odot \approx 4 \times 10^{26}$ W is the luminosity of the sun. Notice that the absolute bolometric magnitude of the sun is equal to 4.72, and that, as luminosities range from $10^{-4}L_\odot$ to $10^6 L_\odot$, bolometric magnitudes decrease from about +15 to $-10$.

### Surface temperature

The effective surface temperature of a star, $T_E$, is defined as the temperature of the black body of the same size which would give the same luminosity. For a star of luminosity $L$ and radius $R$

$$L = 4\pi R^2 \sigma T_E^4, \tag{1.43}$$

where $\sigma$ is Stefan's constant. For the sun $T_E \approx 6000$ K.

The surface temperature can be deduced from a spectral analysis of the radiation from the star. In particular, if the star acts like a black body radiator, the colour of the radiation would give a precise indication of the temperature; the spectrum would peak at a frequency given by $h\nu = 2.82\, kT_E$. In practice one must take into account that the star is not a perfect black body in deducing the temperature from the colour.

An additional source of information on the surface temperature is provided by absorption lines in the spectrum. As radiation passes through the photosphere, the surface region from which most of the observed radiation originates, certain frequencies are absorbed by particular ions and atoms to give a spectrum containing dark absorption lines. The absorption lines in the spectrum permit a classification of stars according to spectral type. The spectral type depends on the degree of excitation and ionization of atoms and ions in the photosphere. It is denoted by a letter O, B, A, F, G, K or M, a sequence which largely reflects a steady decrease in surface temperature from 30,000 K to 3000 K. The sequence is remembered by a mnemonic which these days is considered sexist.

### Luminosity and surface temperature

The main observational properties of a star, its luminosity and its surface temperature, are not uncorrelated. The correlation is usually illustrated in a two-dimensional plot called an Hertzsprung–Russell diagram in which the vertical axis represents the luminosity and the horizontal axis represents the surface temperature; for historical reasons the temperature decreases to the right. When stars are

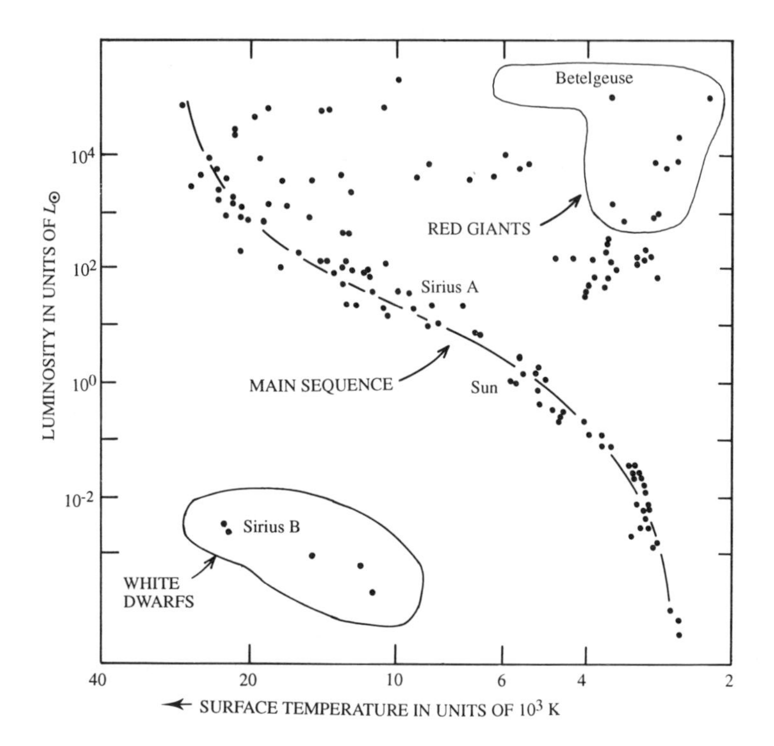


Fig. 1.6 A schematic Hertzsprung–Russell diagram. This diagram provides a snapshot of the luminosity and surface temperature of stars at different stages of their evolution. Most of the observed stars are grouped along a band called the main sequence; these are hydrogen burning stars like the sun. As stars evolve the contraction of the central core is accompanied by an expansion of the outer layers of the star to form luminous stars with low surface temperature; e.g. red giants. The end-point of stellar evolution of a star with a mass comparable to the sun is a compact object supported by degenerate electrons, a white dwarf. The evolution of a more massive star can lead to the formation of a neutron star or a black hole.

represented by a point with coordinates $(T_E, L)$ on this diagram, certain regions are more densely populated than others. The different regions of the H–R diagram are illustrated in Fig. 1.6.

In interpreting the H–R diagram it is important to remember that star formation and evolution is an on-going process. An H–R diagram for stars in a particular region of space provides a snapshot of stars at different stages of their evolution.

As stars evolve they spend most of life burning hydrogen. Hence hydrogen burning stars, like the sun, should give rise to a densely populated region of the H–R diagram. This region is called the *main sequence*. About 80 to 90% of observed stars are main sequence stars.

The relation between the mass and luminosity of a hydrogen burning star can be used to deduce the mass of a star from its position on the main sequence. We recall from Section 1.5 and Fig. 1.5, that the luminosity of a star of mass $M$ is proportional to $M^{\alpha}$, where $\alpha$ is between 3 and 3.5. This mass–luminosity relation can be used to show that the masses range from about $50M_{\odot}$ at the upper left side of the main sequence to about $0.1M_{\odot}$ at the lower right. This result is particularly useful because there is no direct method of measuring the mass of an isolated star; i.e. one that does not belong to a binary or multiple system.

A star does not evolve along the main sequence. It evolves onto the main sequence when a protostar contracts and ignites hydrogen. The star evolves off the main sequence, and moves into the red giant region of the H–R diagram, when the hydrogen in the central nuclear burning region is depleted. Theoretical models indicate that hydrogen burning in the core ceases but continues in a thin shell which moves outwards. The core contracts and heats up, but the outer layers expand to form a star of high luminosity and low surface temperature. We can use Eq. (1.43) to show that a star in the red giant region of the H–R diagram with $L = 1000L_{\odot}$ and $T_E = 4000$ K has a radius of about $70R_{\odot}$. Such a star will stand out conspicuously in the sky. A famous example is Betelgeuse in the constellation Orion.

As the temperature and density at the centre of the star increase, helium and subsequently other nuclear fuels are ignited, but the number of nuclear burning stages depends on the mass of the star. Stars in advanced stages of nuclear burning occupy the top, right hand region of the H–R diagram. Since the time scale for these stages is brief, this region of the H–R diagram is not densely occupied.

Observations indicate that intermediate mass stars end their life by shedding their outer layers to form a planetary nebula which merges with the interstellar medium to leave a remnant with low luminosity and high surface temperature in the white dwarf region of the H–R diagram. The best known white dwarf is Sirius B, which forms a binary system with the bright main sequence star, Sirius A. In fact, the existence of Sirius B was postulated by Bessel in 1834 in order to explain the fact that Sirius A appeared to wobble in the sky; it was later observed to be a star with a low luminosity and high surface temperature. A typical white dwarf has a luminosity of $L = L_{\odot}/100$ and surface temperature $T_E = 16{,}000$ K, and hence a radius given by Eq. (1.43) of about $R_{\odot}/70$. Such a star shines merely because it evolved from the hot core of a red giant. In time it will cool and fade away; see Section 3.4 of Chapter 3.

The Hertzsprung–Russell diagram is of great practical and historical significance in astronomy. In particular, H–R diagrams for different types of star clusters provided the observational framework for the development of our understanding of stellar evolution. Accordingly, H–R diagrams are extensively discussed in books which emphasize the link between observations and theoretical calculations of stellar evolution.

## SUMMARY

### Big bang nucleosynthesis

- Nuclear reactions in the early universe led to a universe in which about 25% of the mass was helium and the remainder mostly hydrogen. This proportion of helium to hydrogen was largely determined by the ratio of neutrons to protons that existed when neutrons and protons ceased to be continually transformed into each other by neutrino reactions described by Eqs. (1.1).

### Gravitational contraction

- Bodies can collapse rapidly if the gravitational energy released is easily absorbed or radiated away. The time for free fall under gravity of a body of uniform density $\rho$ is

$$t_{FF} = \left[\frac{3\pi}{32G\rho}\right]^{1/2} \tag{1.4}$$

- The pressure gradient needed for hydrostatic equilibrium is given by

$$\frac{\mathrm{d}P}{\mathrm{d}r} = -\frac{Gm(r)\rho(r)}{r^2}. \tag{1.5}$$

- The average pressure needed to support a system with gravitational energy $E_{GR}$ and volume $V$ is given by

$$\langle P \rangle = -\frac{1}{3}\frac{E_{GR}}{V}. \tag{1.7}$$

- The internal kinetic energy and gravitational energy of a gas of non-relativistic particles in hydrostatic equilibrium are related by

$$2E_{KE} + E_{GR} = 0. \tag{1.11}$$

  The most important consequence of this relation is that, as a self-gravitating system loses energy, its gravitational energy decreases and its internal kinetic energy increases. Indeed, half the gravitational energy released supplies the energy loss and the other half is used to increase the kinetic energy.

- The corresponding relation for an ultra-relativistic gas is given by

$$E_{KE} + E_{GR} = 0. \tag{1.13}$$

This relation implies that hydrostatic equilibrium becomes precarious as the constituents of the system become ultra-relativistic.

## Star formation

- A gas of mass $M$ consisting of particles of average mass $\overline{m}$ at a temperature $T$ is gravitationally bound if its average density exceeds a critical value given by

$$\rho_J = \frac{3}{4\pi M^2}\left[\frac{3kT}{2G\overline{m}}\right]^3. \tag{1.19}$$

- The temperature of a contracting body ceases to rise when the electrons become degenerate. The maximum temperature attained by a contracting body of mass $M$ is approximately given by

$$kT \approx \left[\frac{G^2\overline{m}^{8/3}m_e}{h^2}\right] M^{4/3}. \tag{1.28}$$

As a result, only bodies with a mass greater than $0.08M_\odot$ can achieve the necessary temperature to ignite hydrogen fusion and become genuine stars. There is also a maximum mass for a normal star which is in the region of 50 to $100M_\odot$. This arises because radiation pressure, a pressure due to ultra-relativistic particles, becomes increasingly important in massive stars and hydrostatic equilibrium becomes precarious.

## The sun

- The mass and radius of the sun are $M_\odot \approx 2 \times 10^{30}$ kg and $R_\odot \approx 7 \times 10^8$ m.
- The average pressure inside the sun is given by

$$\langle P \rangle = -\frac{1}{3}\frac{E_{GR}}{V} \approx \frac{GM_\odot^2}{4\pi R_\odot^4} \approx 10^{14} \text{ Pa}. \tag{1.29}$$

- The typical internal temperature inside the sun is given by

$$kT_I \approx \frac{GM_\odot\overline{m}}{3R_\odot} \approx 0.5 \text{ keV} \quad \text{or} \quad T_I \approx 6 \times 10^6 \text{ K}. \tag{1.31}$$

- The effective surface temperature, $T_E = 6000$ K, is three orders of magnitude less than the typical internal temperature.

- The luminosity of the sun is $L_\odot \approx 4 \times 10^{26}$ W and is approximately given by

$$L_\odot \approx \frac{(4\pi)^2}{3^5} \frac{\sigma}{k^4} G^4 \overline{m}^4 \langle \rho \rangle l M_\odot^3, \tag{1.36}$$

where $l$ is the effective mean free path for radiative diffusion in the sun.

- The luminosity fixes the rate of thermonuclear fusion within the sun. Hydrogen is fused to helium by the reactions of the proton–proton chain, and the dominant reactions are given by

$$p + p \rightarrow d + e^+ + \nu_e, \tag{1.37}$$

$$p + d \rightarrow {}^3\text{He} + \gamma, \tag{1.38}$$

$${}^3\text{He} + {}^3\text{He} \rightarrow {}^4\text{He} + p + p. \tag{1.39}$$

- A solar luminosity of $4 \times 10^{26}$ W implies that about $4 \times 10^{38}$ protons are consumed per second and that at least $2 \times 10^{38}$ neutrinos are radiated every second.

### Stellar nucleosynthesis

- The observed chemical elements in the solar system are largely a product of nucleosynthesis during the big bang and nucleosynthesis during stellar evolution.
- The extent of thermonuclear fusion in a star is determined by its mass. Hydrogen is burnt if the mass is above $0.08M_\odot$. This will be followed by helium burning if the mass is above $0.5M_\odot$, or thereabouts. Stars with masses roughly between 8 and $11M_\odot$ will progress to carbon burning. Every stage of nuclear fusion up to the synthesis of elements near iron occurs in stars with a mass greater than $11M_\odot$.
- Neutron capture in the final stages of stellar evolution leads to the formation of elements heavier than iron.

### Stellar life cycles

- The mass of a star determines the rate of its evolution, and its ultimate fate. Stars like the sun evolve slowly and end their life as white dwarfs. Massive stars evolve rapidly and end their life with a castastrophic collapse when the mass of the central core of iron exceeds the Chandrasekhar limit of about $1.4M_\odot$. The outer layers of the star can be ejected as a supernova and the remaining matter forms a neutron star or black hole. The interstellar medium is enriched with heavier elements by mass loss from evolving stars; the formation of planetary nebulae by intermediate mass stars and the supernova of massive stars are of particular importance in this regard.

### The Hertzsprung–Russell diagram

- The H-R diagram displays the two basic observational properties of a star, its luminosity and its surface temperature. Certain regions of the H–R diagram correspond to stars at particular stages of their evolution; e.g. hydrogen burning stars are on the main sequence. The H–R diagram has played a key role in the link between observations and theoretical calculations of stellar evolution.

## PROBLEMS 1

1.1 Consider a sphere of mass $M$ and radius $R$. Calculate the gravitational potential energy of the sphere assuming (a) a density which is independent of the distance from the centre, and (b) a density which increases towards the centre according to

$$\rho(r) = \rho_C(1 - r/R).$$

In both cases, (a) and (b), write down the average internal pressure needed for hydrostatic equilibrium, and determine how the pressure within the sphere depends on the distance from the centre.

1.2 The globular cluster M13 in Hercules contains about 0.5 million stars with an average mass of about half the solar mass. Use the Jeans criteria (1.19) to check whether this cluster could have been formed in the early universe just after the time when the universe was cool enough for the electrons and nuclei to form neutral atoms; at this time the density of the universe was $\rho \approx 10^{-27}$ kg m$^{-3}$ and the temperature was $T \approx 10^4$ K.

1.3 As the sun evolved towards the main sequence, it contracted under gravity whilst remaining close to hydrostatic equilibrium, and its internal temperature changed from about 30,000 K, Eq. (1.23), to about $6 \times 10^6$ K, Eq. (1.31). (This stage of stellar evolution is called the Kelvin–Helmholtz stage.) Find the total energy radiated during this contraction. Assume that the luminosity during this contraction is comparable to the present luminosity of the sun and estimate the time taken to reach the main sequence.

1.4 The main sequence of the Pleiades cluster of stars consists of stars with mass less than $6\ M_\odot$; the more massive stars have already evolved off the main sequence. Estimate the age of the Pleiades cluster.

1.5 The binding energy per nucleon for $^{56}$Fe is 8.8 MeV per nucleon. Estimate the energy released per kilogram of matter by the sequence of reactions which fuse hydrogen to iron.

1.6 Given that the luminosity of the sun is $4 \times 10^{26}$ W and that the absolute bolometric magnitude of the sun is $M_{BOL} = 4.72$, estimate the distance at which the sun could just be seen by the naked eye. (The naked eye can detect a star of apparent magnitude 6.) Estimate the number of photons incident on the eye per second in this situation.

1.7 Useful bounds can be set on the pressure at a centre of a star without detailed stellar structure calculations. Consider a star of mass $M$ and radius $R$. Let $P(r)$ be the pressure at distance $r$ from the centre and $m(r)$ be the mass enclosed by a sphere of radius $r$. Show that in hydrostatic equilibrium the function

$$P(r) + Gm(r)^2/8\pi r^4$$

decreases with $r$. Hence show that the central pressure satisfies the inequality

$$P_c > \frac{1}{6}\left[\frac{4\pi}{3}\right]^{1/3} G\langle\rho\rangle^{4/3}M^{2/3},$$

where $\langle\rho\rangle$ is the average density.

If you assume that the density $\rho(r)$ decreases with $r$, it is possible to derive a tighter lower bound and, in addition, a useful upper bound for the central pressure. Show that

$$P_c > \frac{1}{2}\left[\frac{4\pi}{3}\right]^{1/3} G\langle\rho\rangle^{4/3}M^{2/3}.$$

In addition, show that

$$P_c < \frac{1}{2}\left[\frac{4\pi}{3}\right]^{1/3} G\rho_C^{4/3}M^{2/3},$$

where $\rho_C$ is the central density.

CHAPTER 2

# Properties of matter and radiation

A stellar interior is an environment in which matter and radiation at high temperature produce a pressure to oppose gravitational contraction. The conditions are extreme: the atoms are ionized, the electrons can be degenerate and ultra-relativistic, and the pressure due to radiation can be significant. Nevertheless, despite this complexity, many of the properties of stellar interiors can be understood by considering the simplest thermodynamic system, the ideal gas. However we shall need to go beyond the familiar ideal gas in which the particles are both classical and non-relativistic, and consider the ideal gas in its most general form. We shall give particular attention to the properties of electron and photon gases and consider their relevance to stellar structure. In addition, we shall consider the thermodynamics of the dissociation of molecules, the ionization of atoms, the photodisintegration of atomic nuclei and the production of particle anti-particle pairs.

## 2.1 THE IDEAL GAS

The ideal gas is a large number of particles occupying quantum states whose energy is unaffected by the interaction between the particles. The particles we have in mind may be atoms, ions, electrons, photons, neutrinos, . . ., etc. The effects of quantum mechanics and special relativity will often be important in the gas; only in particular circumstances will it be an appropriate approximation to treat the particles as classical and non-relativistic.

### Density of states

The gas particles can act like waves and we can use these wave-like properties to enumerate the possible quantum states that can be occupied by the particles. We assume that the particles are confined in a cubical box of volume $V = L^3$. Confinement to such a box implies that the quantum states can be represented by standing waves of the form $\sin k_x x \ \sin k_y y \ \sin k_z z$ with wave vector $\mathbf{k}$ given by

$$\mathbf{k} = (k_x,\ k_y,\ k_z) = (n_x,\ n_y,\ n_z)\frac{\pi}{L}, \tag{2.1}$$

where $n_x$,$n_y$,$n_z$ are positive integers; in other words, an integer number of half-wavelengths can be accommodated between opposite faces of the box.

The quantum numbers $n_x$,$n_y$,$n_z$ can be used to count the quantum states with different wave vectors. For example, $(L/\pi)\mathrm{d}k_x$ distinct values of $n_x$ are encountered if $k_x$ is increased to $k_x + \mathrm{d}k_x$. Hence the number of quantum states with a wave vector $\mathbf{k}$ with components between $k_x$ and $k_x + \mathrm{d}k_x$, $k_y$ and $k_y + \mathrm{d}k_y$, $k_z$ and $k_z + \mathrm{d}k_z$ is

$$\left[\frac{L}{\pi}\right]^3 \mathrm{d}k_x\ \mathrm{d}k_y\ \mathrm{d}k_z.$$

This result can be interpreted geometrically by thinking of a *k-space* defined by positive coordinates $k_x$, $k_y$, $k_z$ in which any volume element contains many quantum states with a density of $[L/\pi]^3$ states per unit volume.

We shall be interested in the quantum states with a wave vector $\mathbf{k}$ with a magnitude between $k$ and $k + \mathrm{d}k$. These states occupy the $k$-space volume (with positive $k_x$, $k_y$, $k_z$) between two spheres of radius $k$ and $k + \mathrm{d}k$. This volume is $4\pi k^2\mathrm{d}k/8$, and the number of states with wave vector magnitude between $k$ and $k + \mathrm{d}k$ is

$$\left[\frac{L}{\pi}\right]^3 \frac{4\pi k^2 \mathrm{d}k}{8}. \tag{2.2}$$

The particle-like properties of these states become apparent if the momentum of the particle is measured. The de Broglie relation, $p = h/\lambda$, implies that if the wave vector has a magnitude $k$ then the momentum has a magnitude $p = \hbar k$, where $\hbar = h/2\pi$. Accordingly, if we set $p = \hbar k$ in Eq. (2.2) we obtain the following expression for the number of quantum states with a momentum with a magnitude between $p$ and $p + \mathrm{d}p$:

$$g(p)\mathrm{d}p = \frac{V}{h^3} 4\pi p^2 \mathrm{d}p.$$

This result has to be modified if the particles have intrinsic angular momentum or spin. For each state with definite momentum, there can be several quantum states corresponding to different orientations of the spin of the particle, or, in other words, to different polarizations of the particle. Thus, when $g_s$ is the number of independent polarizations of the particle, the number of quantum states with a momentum with a magnitude between $p$ and $p + \mathrm{d}p$ becomes

$$g(p)\mathrm{d}p = g_s \frac{V}{h^3} 4\pi p^2 \mathrm{d}p. \tag{2.3}$$

We note that protons, neutrons and electrons are spin $\frac{1}{2}$ particles with $g_s = 2$. Neutrinos also have spin $\frac{1}{2}$ but have only one polarization; i.e. $g_s = 1$. Photons have spin 1 and $g_s = 2$, corresponding to the two independent polarizations of an electromagnetic wave.

### Internal energy

The internal kinetic energy of the gas depends on three factors: the density of states, the energy of each quantum state, and the number of particles in each state. The density of states, $g(p)\mathrm{d}p$, is given by Eq. (2.3). The energy, $\epsilon_p$, of a particle of mass $m$ in a quantum state with momentum $p$ is, according to special relativity, given by

$$\epsilon_p^2 = p^2c^2 + m^2c^4. \tag{2.4}$$

If we represent the average number of particles in a state with energy $\epsilon_p$ by $f(\epsilon_p)$, we can write the internal energy of the gas as

$$E = \int_0^\infty \epsilon_p \, f(\epsilon_p) \, g(p)\mathrm{d}p. \tag{2.5}$$

Similarly, the total number of particles in the gas is

$$N = \int_0^\infty f(\epsilon_p) \, g(p)\mathrm{d}p. \tag{2.6}$$

The macroscopic, thermodynamic properties of the gas may be described by $T$, $P$ and $\mu$, the temperature, pressure and chemical potential. In particular, these parameters determine how the internal energy of the gas is changed by a transfer of heat or entropy, by a compression or expansion, and by a transfer of particles; if the entropy changes by $\mathrm{d}S$, the volume by $\mathrm{d}V$ and the number of particles by $\mathrm{d}N$, then

$$\mathrm{d}E = T\mathrm{d}S - P\mathrm{d}V + \mu\mathrm{d}N. \tag{2.7}$$

The approach to thermodynamic equilibrium is characterized by processes which lead to a uniform temperature, pressure and chemical potential, which, when equilibrium is established, are related by an equation of state.

From the microscopic viewpoint, the temperature, pressure and chemical potential determine the equilibrium distribution of the particles in the quantum states. This distribution depends on whether the particles are identical fermions or bosons:

- Identical fermions obey Fermi–Dirac statistics in which the occupation of states is restricted by the Pauli exclusion principle; not more than one particle can be in a given quantum state and the average number is

$$f(\epsilon_p) = \frac{1}{\exp[(\epsilon_p - \mu)/kT] + 1}. \tag{2.8}$$

- Identical bosons obey Bose–Einstein statistics in which any number of particles may be in a given quantum state. The average number is

$$f(\epsilon_p) = \frac{1}{\exp[(\epsilon_p - \mu)/kT] - 1}. \tag{2.9}$$

These distribution functions are illustrated in Fig. 2.1, which shows that the average occupation of every state decreases as the chemical potential decreases or as the temperature increases. As this happens, the fermion and the boson distribution functions approach the same distribution function, a distribution which is appropriate for a dilute classical gas. In a dilute classical gas, even the states of lowest energy with $\epsilon_p = mc^2$ are scarcely occupied. Indeed, the occupation of such states becomes very much less than one when

$$\exp[(mc^2 - \mu)/kT] >> 1. \tag{2.10}$$

In this case, the +1 in the fermion distribution function, Eq. (2.8), and the $-1$ in the boson distribution function, Eq. (2.9), can be neglected, and in both cases the average number of particles in a quantum state becomes

$$f(\epsilon_p) \approx \exp[-(\epsilon_p - \mu)/kT] << 1. \tag{2.11}$$

It follows that a gas of bosons and a gas of fermions have similar properties when the occupation of every quantum state is low. The fact that there can be at most one fermion but any number of bosons in a given state is of no relevance since the average occupation of any state is very much less than one. Furthermore, dilute gases act as classical systems of particles because the separation between the

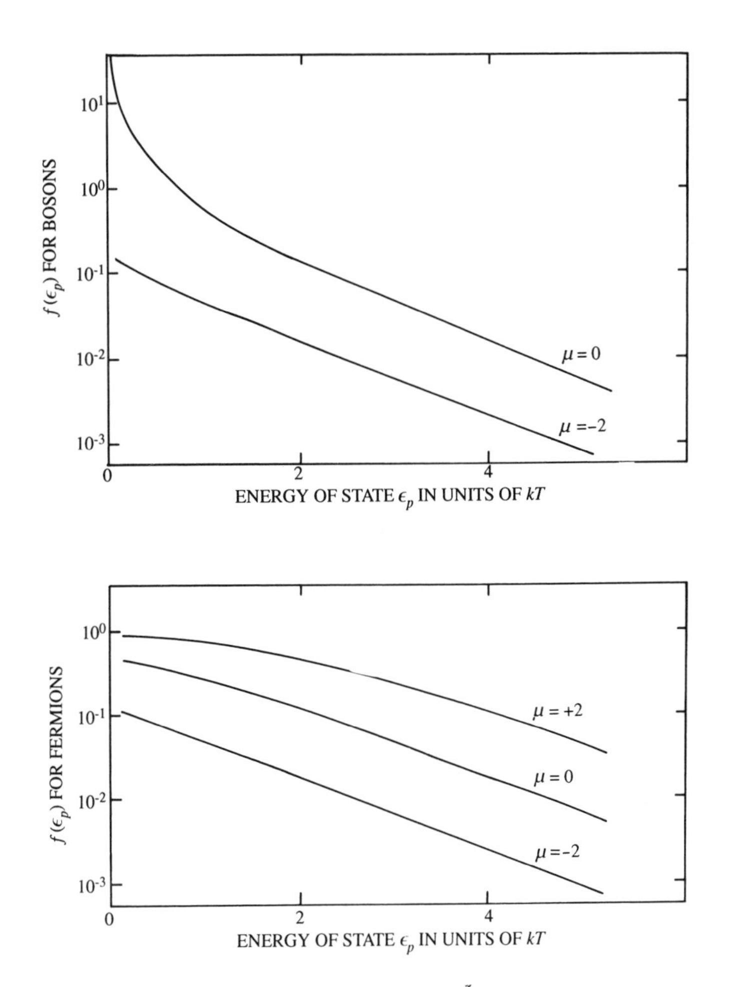


Fig. 2.1 The average number of bosons and fermions in a quantum state with energy $\epsilon_p$ for different values of the chemical potential. The energy scale has been fixed by setting $kT = 1$; e.g. the average occupation of a state at $\mu = -2$ and $\epsilon_p = 2$ on the graph represents the average occupation at $\mu = -2kT$ and $\epsilon_p = 2kT$.

particles allows an observer to keep track of their motion and distinguish particles which are really identical. In particular, there are no quantum effects arising from the identity of the particles. In this situation bosons and fermions obey classical, Maxwell–Boltzmann statistics; they behave like *maxwellions*. We shall see that the quantum and classical behaviour of electron gases play an essential role in stellar physics.

## Pressure in an ideal gas

In Section 1.2 we used classical arguments to derive relations between the pressure and internal energy density of an ideal gas. These were used to show that the hydrostatic equilibrium of a self-gravitating gas becomes more precarious as the gas particles become more relativistic. Here we shall confirm that these relations are also valid when quantum physics is appropriate.

From the fundamental thermodynamic relation, Eq. (2.7), we note that the change in the internal energy brought about by a volume change $dV$ at constant entropy and particle number is equal to $-PdV$, the work done on a system. In such a process, the number of particles in each quantum state remains constant, and the internal energy changes because the energy of each quantum state depends upon the volume. Thus, by using Eq. (2.5), we find that pressure in an ideal gas is given by

$$P = -\frac{\partial E}{\partial V} = -\int_0^\infty \frac{d\epsilon_p}{dV} f(\epsilon_p)\, g(p) dp. \tag{2.12}$$

In order to find the dependence of the quantum state energy on the volume confining the particles, we consider $\epsilon_p$ as a function of $p$, and $p$ as a function of $V$, and write

$$\frac{d\epsilon_p}{dV} = \frac{d\epsilon_p}{dp}\frac{dp}{dV}.$$

According to Eq. (2.1), the wave vector, and hence the momentum, is inversely proportional to $L$. Since $V = L^3$, $p \propto V^{-1/3}$ and

$$\frac{dp}{dV} = -\frac{p}{3V}.$$

Also, the relation (2.4) between energy and momentum gives

$$\frac{d\epsilon_p}{dp} = \frac{pc^2}{\epsilon_p} = v_p,$$

where $v_p$ is the speed of a particle with momentum $p$. Hence we find

$$\frac{d\epsilon_p}{dV} = -\frac{pv_p}{3V}.$$

Substitution of this result into Eq. (2.12) gives the following expression for the pressure in an ideal gas:

$$P = \frac{1}{3V}\int_0^\infty pv_p\, f(\epsilon_p)\, g(p) dp = \frac{N}{3V}\langle pv_p\rangle, \tag{2.13}$$

where the brackets, $\langle\ \rangle$, denote an average over the $N$ particles in the gas. We note that Eq. (2.13) agrees with Eq. (1.8) which was derived using classical physics.

It is easy to show that, when the gas particles are non-relativistic or ultra-relativistic, the pressure is directly proportional to the kinetic energy density of the gas. For non-relativistic particles $\epsilon_p = mc^2 + p^2/2m$ and $v_p = p/m$, and the pressure is

$$P = \frac{2N}{3V}\langle\frac{p^2}{2m}\rangle = \frac{2}{3} \text{ of kinetic energy density.} \tag{2.14}$$

For ultra-relativistic particles $\epsilon_p = pc$ and $v_p = c$, and the pressure is

$$P = \frac{N}{3V}\langle pc\rangle = \frac{1}{3} \text{ of kinetic energy density.} \tag{2.15}$$

We emphasize that these expressions for the pressure are applicable to an ideal gas in its most general form. It is immaterial whether the gas particles are bosons or fermions, or whether they form a dense gas where quantum effects are important or a dilute gas where classical physics is appropriate.

### The ideal classical gas

The reader will be familiar with many of the properties of an ideal classical gas. Our purpose here is twofold: to focus on the effects of relativistic kinematics and to understand when the gas particles are no longer described by classical mechanics. We begin by showing that the familiar equation of state for a classical ideal gas is valid even when the particles are relativistic.

A gas is classical when the average occupation of any quantum state is small and given by Eq. (2.11). The pressure in such a gas can be found by the substitution of Eqs. (2.3) and (2.11) into Eq. (2.13) to give

$$P = \frac{1}{3V}\exp[\mu/kT]\int_0^\infty pv_p \exp[-\epsilon_p/kT]\, g_s\frac{V}{h^3}4\pi p^2 \mathrm{d}p. \tag{2.16}$$

We now use the relativistic relation between energy and momentum, Eq. (2.4), to obtain $\mathrm{d}\epsilon_p = v_p\mathrm{d}p$ and rewrite the integral in Eq. (2.16) as follows

$$\int_0^\infty p^3 \exp[-\epsilon_p/kT]v_p\mathrm{d}p = -kT\int_0^\infty p^3\mathrm{d}(\exp[-\epsilon_p/kT]).$$

Integration by parts then gives

$$\int_0^\infty p^3 \exp[-\epsilon_p/kT]v_p\mathrm{d}p = 3kT\int_0^\infty \exp[-\epsilon_p/kT]\, p^2\mathrm{d}p.$$

Substitution of this result into Eq. (2.16) gives the following expression for the pressure in an ideal classical gas:

$$P = \frac{kT}{V}\exp[\mu/kT]\int_0^\infty \exp[-\epsilon_p/kT]\, g_s\frac{V}{h^3}4\pi p^2 \mathrm{d}p. \tag{2.17}$$

We now compare this expression with the equation for the total number of particles in the gas, Eq. (2.6), which for a classical gas has the form

$$N = \exp[\mu/kT]\int_0^\infty \exp[-\epsilon_p/kT]\, g_s\frac{V}{h^3}4\pi p^2 \mathrm{d}p. \tag{2.18}$$

This comparison leads directly to the equation of state

$$P = \frac{N}{V}kT = nkT. \tag{2.19}$$

We emphasize that this equation of state is valid for classical particles even when they are relativistic. However, as the particles become more energetic there will be additional contributions to the pressure due to particle production. The equation of state can be usefully compared with the relations between the pressure and the kinetic energy density, Eq. (2.14) and Eq. (2.15). We find that the average kinetic energy of a particle in a classical gas is equal to $\frac{3}{2}kT$ if it is non-relativistic and equal to $3kT$ if it is ultra-relativistic.

We now turn to the condition for the particles to form a classical gas. The gas is classical if the average occupation of every quantum state is small compared with unity. This will be the case if the chemical potential satisfies the inequality given by Eq. (2.10). This inequality can be cast in a more useful form by deriving an explicit expression for the chemical potential of a classical gas.

The chemical potential for a classical gas of non-relativistic particles can be found by substituting $\epsilon_p = mc^2 + p^2/2m$ into the expression for the total number of particles in a gas, Eq. (2.18). Integration then gives

$$N = \exp[(\mu - mc^2)/kT]\, g_s\frac{V}{h^3}(2\pi mkT)^{3/2}. \tag{2.20}$$

This may be rearranged to give

$$\mu - mc^2 = -kT\ln\left[\frac{g_s n_Q}{n}\right], \tag{2.21}$$

where $n$ is $N/V$, the density of particles in the gas, and $n_Q$ is defined by

$$n_Q = \left[\frac{2\pi mkT}{h^2}\right]^{3/2} \tag{2.22}$$

We shall see that $n_Q$ is an important parameter in statistical physics; it is called the quantum concentration.

A similar calculation can be carried out for a classical gas of ultra-relativistic particles. If we neglect the rest energy, $mc^2$, of the particles and substitute $\epsilon_p = pc$ into Eq. (2.18), we obtain the chemical potential

$$\mu = -kT \ln \left[ \frac{g_s n_Q}{n} \right], \tag{2.23}$$

where the quantum concentration is now given by

$$n_Q = 8\pi \left[ \frac{kT}{hc} \right]^3. \tag{2.24}$$

Where necessary, we shall distinguish between the quantum concentrations for a non-relativistic gas and an ultra-relativistic gas by using the notation $n_{QNR}$ and $n_{QUR}$.

These expressions for the chemical potential can be used to reveal the physical significance of the inequality (2.10), the condition for a gas to be classical. We see immediately that the inequality,

$$\exp[(mc^2 - \mu)/kT] >> 1$$

is satisfied if the actual particle concentration $n$ is small compared with the quantum concentration $n_Q$. It is easy to see that this is just a more precise way saying that the average separation of the gas particles is large compared with their typical de Broglie wavelength. For non-relativistic particles, $\lambda = h/p \approx h/(mkT)^{1/2}$ and the particle separation is large compared with $\lambda$ if $n << [mkT/h^2]^{3/2} \approx n_{QNR}$. For ultra-relativistic particles, $\lambda = h/p \approx hc/kT$ and their separation is large if $n << [kT/hc]^3 \approx n_{QUR}$. In the simplest terms, gas particles are only classical if their de Broglie wavelengths are small. We shall see in Section 2.3 that photons never form a classical gas because the chemical potential for photons is fixed and equal to zero. However, gases formed from particles with mass, such as electrons and ions, can behave classically or quantum mechanically, depending on their density. The required density for the breakdown of classical physics is lower in a gas of light particles because lighter particles have longer de Broglie wavelengths. Thus, as a star contracts and as its density increases, the electrons, the lightest particles in the ionized interiors, are the first to exhibit the breakdown of classical physics. Electrons are the first to form a quantum gas. Many aspects of stellar structure are affected by this quantum mechanical behaviour.

## 2.2 ELECTRONS IN STARS

As stellar matter is compressed, electrons are the first particles to change their role. They initially form a dilute classical gas and then a dense quantum gas.

Moreover they become increasingly relativistic as the density increases. We can illustrate this trend by considering the electrons in the sun.

In Chapter 1 we assumed that the sun was a body with an average density $1.4 \times 10^3$ kg m$^{-3}$ composed of electrons and ions which form an ideal gas of non-relativistic, classical particles. By considering the pressure needed to support the sun, we found that the typical temperature of this classical gas is $6 \times 10^6$ K. It is easy to confirm that the electrons in the sun are indeed non-relativistic and classical. First, the electrons are non-relativistic because the typical thermal energy, $kT$, is small compared with the rest energy of the electron; in fact, $kT \approx 10^{-3}\ mc^2$. Second, an average solar density of $1.4 \times 10^3$ kg m$^{-3}$ implies an average electron concentration $n$ of about $6 \times 10^{29}$ m$^{-3}$. This should be compared with the quantum concentration, $n_Q$, for electrons at a temperature $6 \times 10^6$ K. Using Eq. (2.22), we find that $n_Q$ is about $3 \times 10^{31}$ m$^{-3}$. Thus, on average the electrons in the sun form a dilute gas with a concentration much less than the quantum concentration. In other words, they form a classical gas. When we focus our attention on the electrons in the central core of the sun, we shall find that even in this dense region the electrons still form a gas which is approximately classical.

Thus at present, the electrons in the sun form a gas of non-relativistic, classical particles. However, it is easy to see that quantum effects will eventually become important when the central core of the sun contracts. According to Eq. (1.11), the thermal kinetic energy increases as the solar core contracts. The typical temperature, $T$, inside a contracting solar core of mass $M$ increases as the radius $R$ of the core decreases. In analogy with Eq. (1.31), we expect

$$kT \approx \frac{GM\overline{m}}{3R}. \tag{2.25}$$

According to Eq. (1.22), the quantum concentration for the electrons in the core increases as the temperature increases; in fact, $n_Q \propto T^{3/2}$. If this temperature dependence is combined with Eq. (2.25), we find that $n_Q \propto R^{-3/2}$. In comparison, the actual concentration of the electrons increases as $n \propto R^{-3}$. Thus, $n$ increases more rapidly than $n_Q$ as the core contracts, and eventually $n$ will exceed $n_Q$. In other words, the process of contraction will lead to an electron gas in which quantum effects are important. Moreover, we shall see that, if this quantum gas becomes more dense, the electrons will become relativistic.

Hence, electrons obey different rules at different times during stellar evolution. We are already familiar with the equation of state for a classical gas of electrons. We now need to find the equation of state for a quantum gas of electrons.

### The degenerate electron gas

Quantum effects dominate when the concentration of electrons becomes large compared with the quantum concentration. This high density requirement can also be viewed as a low temperature requirement. In particular, $n >> n_Q$ is equivalent to $kT << h^2 n^{2/3}/2\pi m$, if $n_Q$ is given by the non-relativistic Eq. (2.22). Thus, a quantum gas is a *cold* gas, but the standard of coldness is set by the density of the gas; a temperature of a billion degrees can be cold in a very dense gas.

A cold gas of electrons is called a degenerate gas because the electrons have fallen into quantum states with the lowest possible energy. Electrons are identical fermions and obey the Pauli exclusion principle. Hence the electrons must be distributed so that each of the quantum states up to a certain energy are occupied fully by one electron and quantum states with higher energy are unoccupied. Such a distribution of electrons must be the zero temperature limit of the Fermi–Dirac distribution (2.8). Indeed, if we set the chemical potential at zero temperature equal to an energy $\epsilon_F$, Eq. (2.8) gives the following for the average number of electrons in a quantum state with energy $\epsilon_p$:

$$f(\epsilon_p) = 1 \quad \text{if} \quad \epsilon_p \leq \epsilon_F, \quad \text{and} \quad f(\epsilon_p) = 0 \quad \text{if} \quad \epsilon_p > \epsilon_F.$$

The energy of the most energetic electrons in a cold electron gas, $\epsilon_F$, is called the Fermi energy; the corresponding momentum, $p_F$, is called the Fermi momentum.

Because every state up to those with a momentum $p_F$ is occupied by one electron and all other states are unoccupied, the total number of electrons in a degenerate gas is the number of states with momentum less $p_F$. Using Eq. (2.3) for the density of states, we find

$$N = \int_0^{p_F} g_s \frac{V}{h^3} 4\pi p^2 \mathrm{d}p = \frac{8\pi V}{3h^3} p_F^3, \tag{2.26}$$

where we have used $g_s = 2$ to account for the two independent spin states of the electron. This equation for the number of electrons may be rearranged to give the Fermi momentum in terms of the electron density,

$$p_F = \left[\frac{3n}{8\pi}\right]^{1/3} h. \tag{2.27}$$

We note that this expression for the Fermi momentum implies that the de Broglie wavelength of the most energetic electrons in a degenerate gas, $\lambda = h/p_F$, is comparable with $n^{-1/3}$, the average distance between the electrons.

The equation of state for a degenerate gas can be found by evaluating the internal energy. We shall consider two special cases corresponding to non-relativistic and ultra-relativistic electrons:

The electrons in a degenerate gas are non-relativistic if $p_F << mc$ which is equivalent to $n << (mc/h)^3$, where $h/mc$ is the Compton wavelength of the electron, $2.4 \times 10^{-12}$ m. In this case the internal energy of the gas can be found by substituting $\epsilon_p = mc^2 + p^2/2m$ into

$$E = \int_0^{p_F} \epsilon_p g_s \frac{V}{h^3} 4\pi p^2 \mathrm{d}p \tag{2.28}$$

to give

$$E = N \left[ mc^2 + \frac{3p_F^2}{10m} \right]. \tag{2.29}$$

We now recall that, according to Eq. (2.14), the pressure in a non-relativistic ideal gas is two-thirds of the kinetic energy density. Hence the pressure in a non-relativistic, degenerate gas is

$$P = n \frac{p_F^2}{5m}. \tag{2.30}$$

The Fermi momentum can be expressed in terms of the electron density using Eq. (2.27), to give an equation of state of the form

$$P = K_{NR} n^{5/3}, \quad \text{where} \quad K_{NR} = \frac{h^2}{5m} \left[ \frac{3}{8\pi} \right]^{2/3}. \tag{2.31}$$

The equation of state takes a different form when the degenerate electrons are predominantly ultra-relativistic. In this case $n >> n_{QUR}$ and $n >> (mc/h)^3$, and an approximate expression for the internal energy can be obtained by substituting $\epsilon_p = pc$ into Eq. (2.28). This gives

$$E = N \frac{3}{4} p_F c \tag{2.32}$$

and the pressure, which now, according to Eq. (2.15), equals one-third of the kinetic energy density, is

$$P = n \frac{1}{4} p_F c. \tag{2.33}$$

The equation of state becomes

$$P = K_{UR} n^{4/3}, \quad \text{where} \quad K_{UR} = \frac{hc}{4} \left[ \frac{3}{8\pi} \right]^{1/3}. \tag{2.34}$$

We note from Eqs. (2.31) and (2.34), that the pressure of a degenerate gas is an increasing function of the density, but the rate of increase becomes less rapid once the particles become ultra-relativistic; i.e. the equation of state becomes less stiff. We shall see in Section 6.1 of Chapter 6 that this has important implications for the stability of white dwarfs.

## A density–temperature diagram

An understanding of when electrons become degenerate and relativistic is important in the theory of stellar evolution. The form of the equation of state changes gradually as the temperature and density vary, and takes on a simple form if the electrons are classical or degenerate, and non-relativistic or ultra-relativistic. For example, the electrons are ultra-relativistic and degenerate with the equation of state $P = K_{UR}n^{4/3}$ if $n >> n_{QUR}$ and $n >> (mc/h)^3$. If the temperature increases and the density decreases the electrons become classical and ultra-relativistic with an equation of state $P = nkT$ when $n << n_{QUR}$ and $kT >> mc^2$.

The classical, quantum, non-relativistic and ultra-relativistic regimes for electrons in an ideal gas are illustrated in Fig. 2.2. To a first approximation, the boundary

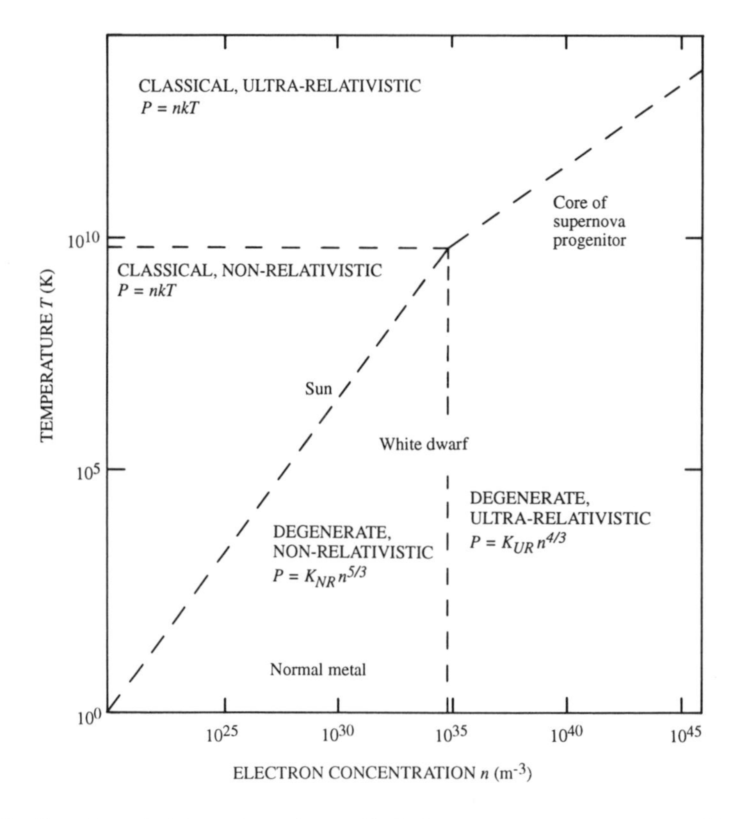


Fig. 2.2 Equation of state regimes for an ideal electron gas at a temperature $T$ and at a density of $n$ electrons per $m^3$. Typical values are shown for the temperature and density for electrons in a normal metal, in the sun, in a white dwarf and in the iron core of an evolved star just prior to a supernova.

lines between the different regimes in this $n - T$ diagram are set by the four equations:

$$n = n_{QNR} \approx 2 \times 10^{21}\ T^{3/2}\ \mathrm{m}^{-3},$$
$$n = n_{QUR} \approx 8 \times 10^{6}\ T^{3}\ \mathrm{m}^{-3},$$
$$n = (mc/h)^3 \approx 7 \times 10^{34}\ \mathrm{m}^{-3},$$
$$T = mc^2/k \approx 6 \times 10^{9}\ \mathrm{K}.$$

Inspection of Fig. 2.2 indicates that dense electron gases are degenerate provided they are not too hot, and that hot electron gases are classical provided they are not too dense.

In practice the electron gas is not ideal because electrons interact. The principal correction to the equation of state usually arises from the electrostatic interaction between electrons and ions. In a classical gas this correction becomes more important as the density increases. This is illustrated by the density dependence of the ratio of the electrostatic energy of interaction to the thermal kinetic energy. This ratio is approximately

$$\frac{E_{ES}}{kT} \approx \frac{Ze^2}{4\pi\epsilon_0 kT}\, n^{1/3} \qquad (2.35)$$

where $Z$ is the charge on the ions and $n^{-1/3}$ is the typical distance between an ion and an electron.

In contrast, the electrostatic correction becomes less important at high density in a degenerate gas. In this case, the typical kinetic energy of the electrons is determined by the Fermi momentum, which in turn depends on the density; see Eq. (2.27). If electrons are non-relativistic, the ratio of the electrostatic energy to the kinetic energy is approximately

$$\frac{E_{ES}}{p_F^2/2m} \approx \frac{Ze^2 n^{1/3}}{4\pi\epsilon_0 p_F^2/2m} \approx \frac{Ze^2 2m}{4\pi\epsilon_0 h^2}\, n^{-1/3}. \qquad (2.36)$$

Thus, as the density of a degenerate gas increases, electrostatic interactions become less important and the ideal gas approximation becomes more appropriate.

### Electrons in the sun

The changing role of electrons in stellar evolution can be illustrated by considering the electrons in the sun. At present, according to theoretical models of the sun, the centre of the sun contains electrons with a concentration of about $8\times10^{31}\ \mathrm{m}^{-3}$ at a temperature of about $1.6\times10^{7}$ K. If we substitute this temperature into Eq. (2.22) we find a quantum concentration for electrons of $1.5 \times 10^{32}\ \mathrm{m}^{-3}$, which is just above the actual concentration. This implies that we can approximately treat the electrons at the centre the sun as classical gas, but that a precise treatment must include small but significant corrections due to degeneracy.

As the sun evolves, the hydrogen content will be reduced and helium will build up in the core. Eventually hydrogen burning will be confined to a shell surrounding a small but growing helium core. Evolutionary models indicate that the temperature will rise to about $2 \times 10^7$ K and the electron concentration to about $3 \times 10^{32}$ m$^{-3}$. The effects of electron degeneracy will now become more important because the quantum concentration at this temperature, $2 \times 10^{32}$ m$^{-3}$, is comparable with the actual concentration.

When the sun evolves away from the main sequence towards the red giant region of the Hertzsprung–Russell diagram (Fig. 1.6), evolutionary models predict it will develop a helium core with a temperature of about $10^8$ K and an electron concentration of $3 \times 10^{34}$ m$^{-3}$. The quantum concentration at this temperature is $2 \times 10^{33}$ m$^{-3}$, which is an order of magnitude less than the actual concentration. Despite a temperature of $10^8$ K, we have a cold, degenerate gas in which most of the electrons occupy the states of lowest energy in accordance with the Pauli exclusion principle.

Even though $10^8$ K is cold enough for the electrons to be degenerate, it is hot enough to ignite the fusion of helium to form carbon. According to Section 1.4, the fusion energy released will cause an increase in the gravitational energy and a decrease in the internal energy. If the core were composed of non-degenerate matter, the fusion control mechanism discussed in Section 1.4 would be operative: if the fusion energy cannot escape, the core will expand and cool, and the rate of fusion will decrease. However, when helium fusion begins in the sun, the core will consist of a classical gas of ions and a degenerate gas of electrons, with the latter providing the bulk of the pressure. The release of excess fusion energy into this material will be accompanied by an expansion and a decrease in the energy of the degenerate electrons but without any appreciable fall in temperature. The rate of fusion will be uncontrolled. Thus, the onset of helium burning in the sun will cause an explosive release of energy in a thermal runaway called a *helium flash*; the peak power could exceed the present luminosity of the sun by a factor of $10^{11}$. But only a fraction of this energy will escape as radiation. Most of it will go into a rapid expansion of the core which lifts the electron degeneracy. Eventually helium fusion will take place in a controlled way in a less dense core of non-degenerate matter.

The sun has insufficient mass to proceed beyond helium burning. It is expected to end its life as an inert white dwarf composed mostly of carbon and oxygen. The mass of this white dwarf is uncertain because of the uncertain mass losses during the red giant and planetary nebula phases of evolution. However, if a white dwarf of $0.5M_\odot$ were formed, the central density will be of the order of $10^9$ kg m$^{-3}$ and the initial temperature, following the completion of helium burning, will be of the order of $10^8$ K. The electrons in the centre of this white dwarf will be degenerate and partially relativistic. In fact, we have an electron gas with concentration $3 \times 10^{35}$ m$^{-3}$, a Fermi momentum of $0.8mc$ and a Fermi energy of $1.3mc^2$; according to Eqs. (2.22) and (2.24), such a gas will be degenerate as long as the temperature is below $10^9$ K. In contrast to the electrons, the carbon ions in the white dwarf are massive particles with small de Broglie wavelengths. They will form a classical system, with an average thermal energy of $\frac{1}{2}kT$ per classical degree of freedom, which will slowly cool down as energy escapes into space. Gravity will tend to compress the ions and the electrons, but the bulk of the resistance will be due to the electrons. Indeed, to a first approximation, the white dwarf is held up by a pressure gradient in a gas of degenerate, partially relativistic electrons.

**Electrons in massive stars**

Electrons play a different role in stars more massive than the sun. First, we note that massive stars can evolve extensively before electron degeneracy affects their evolution. To understand this, we recall from Eq. (2.25) that the typical internal temperature of a stellar core of mass $M$ and radius $R$ supported by a classical ionized gas is given by

$$kT \approx \frac{GM\overline{m}}{3R} \propto M^{2/3}\rho^{1/3}.$$

This equation implies that the temperature rises as a star contracts, but that a given temperature is reached at a lower density $\rho$ if the mass $M$ is higher. It follows that electrons are less likely to be degenerate when nuclear fuels are ignited in a massive star, because the ignition takes place at a lower density. In particular, theoretical models indicate that a star with a mass greater than $11M_\odot$, will normally evolve through all the stages of thermonuclear burning with no effects due to electron degeneracy.

However, electron degeneracy plays a spectacular role at the end of the evolution of a very massive star. Eventually a core of iron is formed. As no further energy can be extracted by nuclear fusion, this core contracts and the electrons become degenerate. The mass of this degenerate core increases as more iron is deposited, and when its mass exceeds the Chandrasekhar mass of about $1.4M_\odot$, it will collapse rapidly. Part of the energy released by this collapse can give rise to the ejection of the outer layers of the star as a supernova. The origin of this instability can be found by considering the electrons in the iron core just before the collapse.

Theoretical models for highly evolved stars suggest that the iron core has a temperature of about $8 \times 10^9$ K and a density of about $4 \times 10^{12}$ kg m$^{-3}$ just before collapse. It is easy to show that the electrons in the core at this stage are degenerate and predominantly ultra-relativistic. First, the electron concentration is approximately $10^{39}$ m$^{-3}$ which implies a relativistic Fermi momentum of about $12mc$. Second, according to Eq. (2.24), the quantum concentration for ultra-relativistic electrons at a temperature $8 \times 10^9$ K is a factor of 1000 smaller than the actual concentration. Thus, we have a gas of degenerate electrons occupying the states of lowest possible energy in accordance with the Pauli exclusion principle. But because the density is so high, most of these electrons are ultra-relativistic; in fact their average energy is $9mc^2$. The collapse of the core is a direct result of the ultra-relativistic nature of the electron gas attempting to support it. As discussed in Section 1.2 of Chapter 1, hydrostatic equilibrium becomes precarious whenever gravity is opposed by the pressure of a gas of ultra-relativistic particles. Indeed, we shall show explicitly in Section 6.1 of Chapter 6, that a gas of ultra-relativistic, degenerate electrons cannot support a mass greater than $1.4M_\odot$.

## 2.3 PHOTONS IN STARS

To a first approximation a star consists of matter and radiation in thermodynamic equilibrium. Indeed, the pressure due to radiation inside a star can be nearly as important as the pressure due to electrons and ions. For this reason we shall review some of the properties of black body radiation, or, in other words, the properties of a photon gas in thermodynamic equilibrium.

### The photon gas

Electromagnetic radiation in equilibrium in a black body cavity can be thought of as an ideal gas of photons. This gas is the simplest ideal gas of all because all the particles move at the same speed, the speed of light. The unusual property of the photon gas is that the number of particles can change; photons are zero mass bosons which can be created and destroyed. We recall that a change in the internal energy of any gas is given by Eq. (2.7),

$$\mathrm{d}E = T\mathrm{d}S - P\mathrm{d}V + \mu \mathrm{d}N,$$

where the third term, involving the chemical potential $\mu$ and the number of particles $N$, describes the effect of a change in particle number. In a photon gas, $N$ is free to change; in particular photons are destroyed or created until equilibrium is established. For example, at fixed energy $E$ and volume $V$ the number of photons changes until the entropy $S$ is a maximum. This equilibrium is characterized by

$$\frac{\partial S}{\partial N} = -\frac{\mu}{T} = 0. \tag{2.37}$$

Similarly, at fixed $T$ and $V$, photons are destroyed or created until the free energy, $F = E - TS$, is a minimum. Since

$$\mathrm{d}F = -S\mathrm{d}T - P\mathrm{d}V + \mu \mathrm{d}N, \tag{2.38}$$

such an equilibrium is characterized by $\partial F/\partial N$ or $\mu$ equal to zero. Thus, a photon gas in equilibrium has zero chemical potential. Its properties may be deduced by setting $\mu = 0$ in the Bose–Einstein distribution function given by Eq. (2.9).

The number of photons in states with momentum between $p$ and $p + \mathrm{d}p$ can be found by using Eq. (2.3) for the density of states and Eq. (2.9) for the distribution function to give

$$N(p)\,\mathrm{d}p = \frac{1}{\exp(\epsilon_p/kT) - 1}\, g_s \frac{V}{h^3} 4\pi p^2 \mathrm{d}p, \tag{2.39}$$

where $\epsilon_p = pc$ and $g_s = 2$ since the photon is a particle with zero mass and two states of polarization. The number of photons per unit volume is

$$n = \frac{1}{V}\int_0^\infty N(p)\mathrm{d}p = 8\pi \left[\frac{kT}{hc}\right]^3 \int_0^\infty \frac{x^2}{e^x - 1}\mathrm{d}x, \tag{2.40}$$

where we have introduced a dimensionless integration variable $x = pc/kT$. Similarly, the energy per unit volume in the photon gas is given by

$$u = \frac{1}{V}\int_0^\infty \epsilon_p\, N(p)\mathrm{d}p = 8\pi \left[\frac{kT}{hc}\right]^3 kT \int_0^\infty \frac{x^3}{e^x - 1}\mathrm{d}x. \tag{2.41}$$

The integrals in Eqs. (2.40) and (2.41) can be related to a special function called the Riemann Zeta Function: The binomial expansion gives

$$\int_0^\infty \frac{x^2}{e^x - 1}\mathrm{d}x = \int_0^\infty x^2[e^{-x} + e^{-2x} + e^{-3x} + \cdots]\mathrm{d}x,$$

and if we integrate each term we obtain

$$\int_0^\infty \frac{x^2}{e^x - 1}\mathrm{d}x = 2\left[\frac{1}{1^3} + \frac{1}{2^3} + \frac{1}{3^3} + \cdots\right] = 2\zeta(3) = 2.404.$$

Similarly

$$\int_0^\infty \frac{x^3}{e^x - 1}\mathrm{d}x = 6\left[\frac{1}{1^4} + \frac{1}{2^4} + \frac{1}{3^4} + \cdots\right] = 6\zeta(4) = \frac{\pi^4}{15}.$$

Thus, Eq. (2.40) for the photon number density and Eq. (2.41) for the photon energy density can be simplified to give

$$n = bT^3, \quad \text{where} \quad b = 2.404 \times \frac{8\pi k^3}{h^3 c^3} = 2.03 \times 10^7\ \mathrm{K^{-3}\ m^{-3}}, \tag{2.42}$$

and

$$u = aT^4, \quad \text{where} \quad a = \frac{8\pi^5 k^4}{15h^3c^3} = 7.565 \times 10^{-16}\ \text{J K}^{-4}\ \text{m}^{-3}. \qquad (2.43)$$

These two equations imply that, $u = 2.70nkT$. Hence the average energy of a photon in a photon gas at temperature $T$ is $2.70\,kT$; the corresponding results for non-relativistic and ultra-relativistic particles in a dilute classical gas are $\frac{3}{2}kT$ and $3kT$, respectively; see the paragraph following Eq. (2.19).

Photons give rise to a pressure called the radiation pressure. Since, according to Eq. (2.15), the pressure due to ultra-relativistic particles is one-third of the kinetic energy density, the radiation pressure at temperature $T$ is

$$P_r = \frac{1}{3}u = \frac{1}{3}aT^4. \qquad (2.44)$$

It is straightforward to determine the relation between the properties of a photon gas and the properties of a black body radiator. A black body radiator at temperature $T$ can be formed by making a small hole in the surface enclosing a photon gas in equilibrium at temperature $T$. Photons will escape like ordinary effusing gas particles at a rate of $nc/4$ per unit area; i.e. on average they move towards and escape from the hole with a speed which is $\frac{1}{4}$ of the actual speed of the particles. Similarly, the rate at which energy escapes is $uc/4$ per unit area. This can be identified with $\sigma T^4$, the power radiated by unit area of a black body, to give a value for Stefan's constant,

$$\sigma = ac/4 = 5.67 \times 10^{-8}\ \text{W K}^{-4}\ \text{m}^{-2}. \qquad (2.45)$$

Photons with all possible energies or frequencies are radiated. Clearly, the intensity radiated at a particular frequency is $c/4$ times the photon energy density at this frequency. If we use Eq. (2.39) and $\epsilon_p = pc = h\nu$, we obtain Planck's formula for the intensity radiated in the frequency range $\nu$ to $\nu + \mathrm{d}\nu$:

$$I_\nu \mathrm{d}\nu = \frac{c}{4}u_\nu \mathrm{d}\nu = \frac{c}{4}\frac{h\nu}{\exp(h\nu/kT) - 1}\frac{8\pi\nu^2}{c^3}\mathrm{d}\nu. \qquad (2.46)$$

This equation implies that the intensity $I_\nu$ and the energy density $u_\nu$ have a maximum at $\nu = 2.82\,kT/h$. In other words, the most probable energy of a photon in radiation at temperature $T$ is $2.82\,kT$, which is slightly higher than the average energy of $2.70\,kT$.

### Radiation pressure in stars

In order to appreciate the quantitative aspects of thermal radiation we list in Table 2.1 the properties of radiation at two particular temperatures, the temperature

TABLE 2.1 The thermal properties of electromagnetic radiation in equilibrium at two temperatures. A temperature of $6 \times 10^3$ K is a typical temperature for the solar surface and $6 \times 10^6$ K is a typical temperature for the solar interior.

| Property | Solar surface at $6 \times 10^3$ K | Solar interior at $6 \times 10^6$ K |
|---|---|---|
| Average photon energy | 1.4 eV | 1.4 keV |
| Photon density, $n$ | $4 \times 10^{18}$ m$^{-3}$ | $4 \times 10^{27}$ m$^{-3}$ |
| Radiation energy density, $u$ | 1 J m$^{-3}$ | $10^{12}$ J m$^{-3}$ |
| Radiation pressure, $P_r$ | 0.33 Pa | $0.33 \times 10^{12}$ Pa |
| Radiation intensity, $\sigma T^4$ | 73 MW m$^{-2}$ | $73 \times 10^{12}$ MW m$^{-2}$ |

of the solar photosphere, $6 \times 10^3$ K, and the typical temperature inside the sun, $6 \times 10^6$ K.

We note that the radiation pressure at the solar surface is tiny by terrestrial standards of pressure, comparable with the pressure exerted by butter on a slice of buttered bread. In the solar interior, the radiation pressure is much greater, more than a million terrestrial atmospheres. Nevertheless, this pressure is much smaller than the pressure needed to support the sun against gravity; according to Eq. (1.29) this is $10^{14}$ Pa. Thus, we were justified in Section 1.4 to neglect radiation and assume that the sun is primarily supported by the pressure generated by electrons and ions.

However it is easy to show that radiation pressure cannot be neglected in stars more massive than the sun. To do so, we recall from Eq. (1.11) that the thermal kinetic energy of a star in hydrostatic equilibrium is related to its gravitational potential energy. This implies that the typical internal temperature $T_I$ in a star of mass $M$ and radius $R$ is approximately proportional to $M/R$. The electrons and ions have densities which are proportional to $M/R^3$, and these particles supply a 'gas' pressure,

$$P_g = n_e kT_I + n_i kT_I \propto \frac{M^2}{R^4}. \tag{2.47}$$

In contrast, the radiation pressure is given by

$$P_r = \frac{1}{3} a T_I^4 \propto \frac{M^4}{R^4}. \tag{2.48}$$

Hence

$$\frac{P_r}{P_g} \propto M^2. \tag{2.49}$$

Thus the ratio of the radiation pressure generated by photons to the 'gas' pressure generated by the electrons and ions increases with the mass of the star. This ratio is small for the sun. But we shall see in Chapter 5 that the radiation pressure becomes comparable with the 'gas' pressure if the mass of the star exceeds $50\,M_\odot$. Furthermore, we recall that hydrostatic equilibrium of a self-gravitating system becomes precarious if the pressure of support is generated by ultra-relativistic particles. Hence radiation pressure is likely to have a destabilizing effect on massive stars.

## 2.4 THE SAHA EQUATION

Molecules are dissociated, atoms are ionized and atomic nuclei are photodisintegrated by radiation. The underlying reaction mechanisms are complex and varied. However, simple and powerful results are readily derived if we assume that matter and radiation are in thermodynamic equilibrium. To illustrate the general ideas, we shall consider hydrogen in equilibrium with radiation and derive the Saha equation for the ionization of atomic hydrogen.

We begin by considering the physical significance of the chemical potential. In a system containing one sort of particle, particles move from a region of high chemical potential to a region of low chemical potential until the chemical potential is the same everywhere. Similarly, if the system consists of particles of type $A$, $B$, $C$ and $D$, which can be transformed into each other via the reactions

$$A + B \rightleftharpoons C + D,$$

thermodynamic equilibrium is reached when the chemical potential of particles $A$ and $B$ equals the chemical potential of particles $C$ and $D$; i.e.

$$\mu(A) + \mu(B) = \mu(C) + \mu(D).$$

These ideas can be applied to the ionization of hydrogen.

The electron in the hydrogen atom can occupy bound states with discrete energies $\epsilon_n$ labelled by the quantum number $n = 1, 2, \cdots$. When the atom is ionized the electron can occupy unbound states with momentum $\mathbf{p}$ and energy $\epsilon_p$. These energy levels are illustrated in Fig. 2.3.

The interaction with photons can cause the hydrogen atom to be excited and ionized. Indeed, at high temperatures we have a dynamic situation in which atoms are continually excited and ionized, and in which electrons are continually captured and atoms are de-excited. If we assume that the atoms, ions, electrons and photons are in thermodynamic equilibrium, we can find the proportion of atoms which are excited and ionized. In particular, the number of hydrogen atoms in states with energy $\epsilon_n$ can be found by considering the dynamic deadlock set up by the reactions

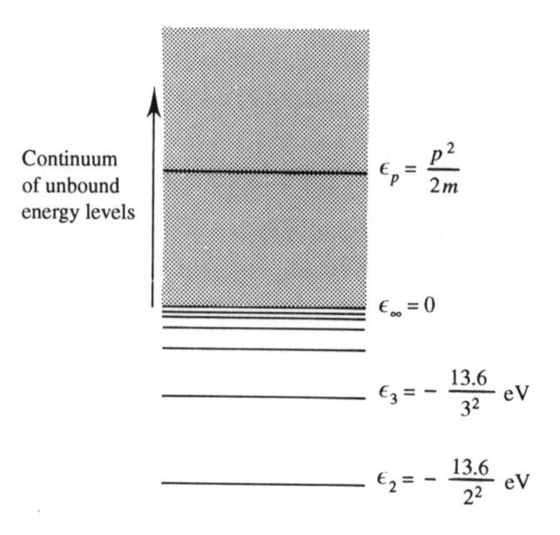


Fig. 2.3 The bound and unbound energy levels of the hydrogen atom.

$$\gamma + \mathrm{H}_n \rightleftharpoons e^- + p. \tag{2.50}$$

Since the chemical potential of the photon is zero, thermodynamic equilibrium is characterized by

$$\mu(\mathrm{H}_n) = \mu(e) + \mu(p). \tag{2.51}$$

If the density is sufficiently low we can assume that the electrons, protons and hydrogen atoms behave as classical particles in an ideal gas, and, if they are non-relativistic, we can use Eq. (2.21) to evaluate the chemical potentials:

$$\mu(e) = m_e c^2 - kT \ln \left[ \frac{g_e n_{Qe}}{n_e} \right], \tag{2.52}$$

$$\mu(p) = m_p c^2 - kT \ln \left[ \frac{g_p n_{Qp}}{n_p} \right], \tag{2.53}$$

$$\mu(\mathrm{H}_n) = m(\mathrm{H}_n) c^2 - kT \ln \left[ \frac{g(\mathrm{H}_n) n_{Qp}}{n(\mathrm{H}_n)} \right]. \tag{2.54}$$

In these equations $n_e$ and $n_p$ are the concentrations of electrons and protons, and $n(\mathrm{H}_n)$ is the concentration of hydrogen atoms in a state with energy $\epsilon_n$; the mass of such an atom $m(\mathrm{H}_n)$ is given by

$$m(\mathrm{H}_n)c^2 = m_e c^2 + m_p c^2 + \epsilon_n.$$

The quantum concentrations are denoted by a subscript $Q$. Because these depend on the mass of the particle, Eq. (2.22), the quantum concentrations for hydrogen atoms and ions are almost identical and they are denoted by $n_{Qp}$. The number of quantum states with a particular energy is denoted by $g$. Since electrons and protons both have spin half, $g_e = 2$ and $g_p = 2$. The number of hydrogen atom states $g(\mathrm{H}_n)$ with energy $\epsilon_n$ is determined by the degeneracy arising from the spin and the relative orbital angular momentum of the electron and proton in the atom. There can be several orbital angular momentum states with the same energy $\epsilon_n$; for example for $n = 2$ there are three p-states and one s-state. In general, the total number of hydrogen atom states with energy $\epsilon_n$ is $g(\mathrm{H}_n) = g_n g_e g_p$ with $g_n = n^2$.

Substituting these chemical potentials into condition (2.51) gives the Saha equation for the equilibrium concentrations of electrons, protons and hydrogen atoms in an ideal gas at temperature $T$. We find

$$\frac{n(\mathrm{H}_n)}{n_e n_p} = \frac{g_n}{n_{Qe}} \exp[-\epsilon_n/kT], \tag{2.55}$$

where the quantum concentration of the electron is

$$n_{Qe} = \left[\frac{2\pi m_e kT}{h^2}\right]^{3/2} \approx 2 \times 10^{21}\ T^{3/2}\ \mathrm{m}^{-3}. \tag{2.56}$$

Because the Saha equation (2.55) is so important, we shall seek a clearer insight into its physical significance by giving an alternative, more intuitive derivation. As already mentioned, the Saha equation describes the result of a dynamic deadlock in which the reaction rate for $\gamma + \mathrm{H}_n \rightarrow e^- + p$ balances the rate for $e^- + p \rightarrow \gamma + \mathrm{H}_n$. Because the rate for the former is proportional to $n(\mathrm{H}_n)$ and the rate for the latter is proportional to $n_e n_p$, it follows that

$$\frac{n(\mathrm{H}_n)}{n_e n_p} = f_n(T),$$

where $f_n(T)$ is some function of the temperature. We expect $f_n(T)$ to be proportional to the probability that an electron is bound, and inversely proportional to the probability that an electron is unbound.

To show that this is indeed the case, we consider an electron and a proton in a box of unit volume at temperature $T$. First, there are $g_e g_n$ bound states with energy $\epsilon_n$ available to the electron. Furthermore, to be in such state the electron has to borrow an energy $\epsilon_n$ from the environment, and the probability of a successful loan is proportional to the Boltzmann factor of $\exp[-\epsilon_n/kT]$. Hence the probability that the electron is bound in a state with energy $\epsilon_n$ is proportional to

$$g_e g_n \exp[-\epsilon_n/kT].$$

Secondly, the density of unbound states available to the electron with momentum between $p$ and $p+\mathrm{d}p$ and kinetic energy $\epsilon_p$ is given by Eq. (2.3), and the probability of acquiring this energy is proportional to $\exp[-\epsilon_p/kT]$. Hence the probability that the electron is unbound with any positive energy is proportional to

$$\int_0^\infty \exp[-\epsilon_p/kT] g_e \frac{1}{h^3} 4\pi p^2 \mathrm{d}p = g_e n_{Qe}.$$

If we assume that the constants of proportionality for these two probabilities are the same, and if we set $f_n(T)$ equal to the ratio of the probabilities, we find that

$$\frac{n(\mathrm{H}_n)}{n_e n_p} = f_n(T) = \frac{g_e g_n \exp[-\epsilon_n/kT]}{g_e n_{Qe}}.$$

This equation is identical to the Saha equation (2.55).

When the electron is bound, it can be in any one of the bound states labelled by the quantum number $n$. The concentration of un-ionized hydrogen atoms is found by summing over all values of $n$. Using the Saha equation (2.55), we find

$$\frac{n(\mathrm{H})}{n_e n_p} = \frac{1}{n_{Qe}} \sum_{n=1}^{n=\infty} g_n \exp[-\epsilon_n/kT]. \tag{2.57}$$

We may rewrite this as

$$\frac{n(\mathrm{H})}{n_e n_p} = \frac{Z}{n_{Qe}} \exp[E_i/kT] \tag{2.58}$$

where $E_i = -\epsilon_1 = 13.6$ eV, the ionization energy of the ground state, and where $Z$ is the function

$$Z = \sum_{n=1}^{n=\infty} g_n \exp[-(\epsilon_n - \epsilon_1)/kT]. \tag{2.59}$$

Note that $\epsilon_n - \epsilon_1$ is the excitation energy of the $n$th state. The function $Z$ is called a partition function. As it stands it is given by a sum which is divergent. But in practice, $Z$ is of the order of unity because the sum is terminated when the value of $n$ corresponds to a state whose spatial extent is comparable with the distance between the gas particles.

Finally we will change the notation slightly and replace $n_p$ by $n(\mathrm{H}^+)$. Then according to Eq. (2.58) the ratio of ionized to un-ionized atoms of hydrogen in a gas at temperature $T$ is approximately

$$\frac{n(\mathrm{H}^+)}{n(\mathrm{H})} \approx \frac{n_{Qe}}{n_e} \exp[-E_i/kT]. \tag{2.60}$$

We note that the degree of ionization depends markedly on the temperature. But it is also inversely proportional to the electron concentration $n_e$. It follows that the ionization increases if the density of the gas decreases. In effect, once atoms are ionized they are less likely to capture an electron if the gas is very dilute.

In order to explore the strong temperature dependence of ionization, we take the logarithm of Eq. (2.60) and obtain

$$\ln\left[\frac{n(\mathrm{H}^+)}{n(\mathrm{H})}\right] = F - \frac{E_i}{kT} \quad \text{where} \quad F = \ln\left[\frac{n_{Qe}}{n_e}\right]. \tag{2.61}$$

Throughout this entire section we have assumed that the electrons form a classical gas with $n_e << n_{Qe}$. Hence $F$ is a positive slowly varying function of temperature. According to Eq. (2.61), the ratio $n(\mathrm{H}^+)/\mathrm{n(H)}$ increases from $e^{-1}$ to $e^{+1}$, i.e. from 0.37 to 2.72, when the temperature increases from $kT = E_i/(F+1)$ to $E_i/(F-1)$. Thus when $F$ is large, as in a very dilute electron gas, the onset of ionization occurs rapidly near $kT = E_i/F$.

## 2.5 IONIZATION IN STARS

In this section we shall consider some of the more important consequences of the ionization of the matter in the interior of stars and in the outer regions of stars.

### Stellar interiors

In order to gain an understanding of the degree of ionization in stellar interiors, we shall, for the sake of simplicity, first consider matter which consists dominantly of hydrogen. The concentrations of the hydrogen atoms, the protons (or $\mathrm{H}^+$ ions), and the electrons will be denoted by $n(\mathrm{H})$, $n(\mathrm{H}^+)$ and $n_e$, respectively. If the mass of the electrons is neglected, the mass density is given by

$$\rho = [n(\mathrm{H}) + n(\mathrm{H}^+)]\, m_{\mathrm{H}}, \tag{2.62}$$

where $m_{\mathrm{H}}$ is the mass of the hydrogen atom. Furthermore, the particle concentrations can be expressed in terms of the density and the fraction of the hydrogen that is ionized. If this fraction is denoted by $x(\mathrm{H})$ then

$$n_e = n(\mathrm{H}^+) = x(\mathrm{H})\rho/m_{\mathrm{H}} \quad \text{and} \quad n(\mathrm{H}) = [1 - x(\mathrm{H})]\rho/m_{\mathrm{H}}. \tag{2.63}$$

The fraction of ionized atoms, $x(\mathrm{H})$, can then be found by substituting Eq. (2.63) into the Saha equation (2.60) and using the ionization energy $E_i = 13.6$ eV. This gives

$$\frac{[1 - x(\mathrm{H})]}{x(\mathrm{H})^2} \approx \frac{\rho/m_{\mathrm{H}}}{10^{21}T^{3/2}} \exp[158{,}000/T], \tag{2.64}$$

where $T$ is the temperature in degrees kelvin. In Section 1.4 we considered a simple model in which the sun was considered to be a globe of ionized gas with an average density of $1.4 \times 10^3$ kg m$^{-3}$ at a temperature of $6 \times 10^6$ K. At this density and temperature, Eq. (2.64) gives $[1 - x(\mathrm{H})]/x(\mathrm{H})^2 \approx 0.055$, indicating that the fraction of hydrogen ionized, $x(\mathrm{H})$, is about 95%. In fact, this calculation underestimates the degree of ionization. Equation (2.64) is not accurate at this density because, even though the electrons and hydrogen ions are small enough to form an ideal gas, the hydrogen atoms are not; their size is comparable with the typical distance between the particles, $d = (\rho/m_{\mathrm{H}})^{-1/3} \approx 10^{-10}$ m. The atoms interact strongly with the gas particles and the likelihood of ionization is increased.

We now assess the extent of the ionization of heavy atoms in the sun. Even though the inner electrons in such atoms are very tightly bound, the ionization is almost complete. This arises largely because small quantities of these atoms are immersed in a dilute electron gas formed by the ionization of hydrogen. To illustrate this, we consider a few carbon atoms in a gas of hydrogen at $T = 6 \times 10^6$ K and $\rho = 1.4 \times 10^3$ kg m$^{-3}$ which is fully ionized to give $n_e \approx \rho/m_{\mathrm{H}} \approx 8 \times 10^{29}$ free electrons per cubic metre. Because the carbon nucleus has charge 6, the ionization energy of the last electron in carbon is $Z^2$ or 36 times the ionization energy of hydrogen. We can adapt Eq. (2.60) to find the approximate ratio of fully ionized carbon atoms to carbon atoms which have only lost 5 electrons. This ratio, which depends on the concentration of electrons provided by the ionization of the hydrogen, is given by

$$\frac{n(6)}{n(5)} \approx \frac{10^{21}T^{3/2}}{n_e} \exp[-36 \times 158{,}000/T] \approx 10. \tag{2.65}$$

This calculation, like the earlier one, underestimates the degree of ionization. However it indicates that, to a first approximation, the atoms inside a star like the sun are completely ionized.

Complete ionization greatly simplifies the analysis of the properties of matter inside stars. In particular, we can find simple expressions for the total number of particles and their average mass: let $X_1$, $X_4$, and $X_A$ be the mass fractions of hydrogen, helium and heavy elements. If the material was not ionized, the number of H and He atoms and the number of heavy atoms per unit volume would be

$$n_1 = X_1\rho/m_{\mathrm{H}}, \quad n_4 = X_4\rho/4m_{\mathrm{H}} \quad \text{and} \quad n_A = X_A\rho/Am_{\mathrm{H}}. \tag{2.66}$$

When ionized, a hydrogen atom yields 2 gas particles, a proton and an electron. A fully ionized helium atom yields 3 particles, a nucleus and two electrons, whereas a fully ionized heavy atom of mass number $A$ and atomic number $Z$ yields a nucleus and $Z$ electrons, about $A/2$ particles in total. Hence the total number of particles per unit volume in a fully ionized gas is

$$n \approx 2n_1 + 3n_4 + \frac{A}{2}n_A = \left[2X_1 + \frac{3}{4}X_4 + \frac{1}{2}X_A\right] \rho/m_{\rm H}.$$

Since $X_1 + X_4 + X_A = 1$, we have

$$n \approx [1 + 3X_1 + 0.5X_4]\rho/2m_{\rm H}. \tag{2.67}$$

Hence the average mass of the gas particles is

$$\overline{m} = \rho/n \approx 2m_{\rm H}/[1 + 3X_1 + 0.5X_4]. \tag{2.68}$$

For example, the Standard Solar Model, Bahcall (1989), assumes that the sun was formed from matter with $X_1 = 0.71, X_4 = 0.27$ and $X_A = 0.02$, which, when ionized, forms a gas of particles with average mass of $\overline{m} \approx 0.61$ amu. The Standard Model predicts that hydrogen burning in the sun has reduced the hydrogen content and increased the helium content so that, at present, the mass fractions in the central regions are approximately $X_1 = 0.34$, $X_4 = 0.64$ and $X_A = 0.02$; this material has average particle mass of 0.85 amu.

Finally, it is also useful to have expressions for the number of electrons and the number of ions per unit volume in a fully ionized gas. It is straightforward to show that these are given by

$$n_e \approx [1 + X_1]\rho/2m_{\rm H} \quad \text{and} \quad n_i \approx [2X_1 + 0.5X_4]\rho/2m_{\rm H}. \tag{2.69}$$

### Stellar atmospheres

We recall from Section 1.7 that stars are classified according to their spectral type. The classification, denoted by a letter O, B, A, F, G, K or M, largely reflects a steady decline in surface temperature from about 30,000 K to about 3000 K. The atoms in stellar atmospheres at these temperatures are partially ionized. Moreover, if the chemical composition, temperature and density of the stellar atmosphere are known, the degree of ionization of the various atomic species may be estimated by applying the Saha equation. In general, the metallic elements like Li, Na, Mg, Al, K, Ca, etc, with an ionization energy of about 5 eV, are predominantly ionized. Elements, such as H, C, N, O, F, P, S, Cl, Ar, which have ionization energies in the range 10 to 20 eV, tend to be partially ionized, whereas He and Ne, noble gas elements with ionization energies above 20 eV, are only partially ionized even in the hottest stellar atmospheres.

The general situation can be understood by considering the ionization of three representative elements, sodium, hydrogen and helium, which have ionization energies of 5.14, 13.6 and 24.6 eV, respectively. If these ionization energies are substituted into the Saha equation we obtain

$$\frac{n(\mathrm{Na}^+)}{n(\mathrm{Na})} \approx \frac{10^{21}T^{3/2}}{n_e} \exp[-60{,}000/T], \tag{2.70}$$

$$\frac{n(\mathrm{H}^+)}{n(\mathrm{H})} \approx \frac{10^{21}T^{3/2}}{n_e} \exp[-158{,}000/T], \tag{2.71}$$

$$\frac{n(\mathrm{He}^+)}{n(\mathrm{He})} \approx \frac{10^{21}T^{3/2}}{n_e} \exp[-286{,}000/T]. \tag{2.72}$$

The exponential factors in these equations give rise to huge differences in the degree of ionization of sodium, hydrogen and helium. For example, at the temperature of the solar photosphere, 6000 K,

$$\frac{n(\mathrm{Na}^+)}{n(\mathrm{Na})} \approx 10^7 \frac{n(\mathrm{H}^+)}{n(\mathrm{H})} \quad \text{and} \quad \frac{n(\mathrm{He}^+)}{n(\mathrm{He})} \approx 10^{-10} \frac{n(\mathrm{H}^+)}{n(\mathrm{H})}. \tag{2.73}$$

The ionization of metallic elements plays a crucial role in stellar atmospheres. We see from Eq. (2.73), that the ionization of sodium is a factor of $10^7$ larger than the ionization of hydrogen. This factor more than compensates for the low abundance of sodium relative to hydrogen in stellar atmospheres; this abundance is about $10^{-6}$ in stars like the sun. Thus, even though stellar matter largely consists of hydrogen and helium with traces of heavier elements, most of the free electrons in stellar atmospheres are due to the ionization of metallic elements like sodium. Moreover, the degree of ionization of other elements, such as hydrogen and helium, depends on the concentration of these electrons. For the solar atmosphere $T$ and $n_e$ are typically of the order of 6000 K and $10^{19}$ m$^{-3}$. If we substitute these values into the Saha equation we obtain

$$\frac{n(\mathrm{H}^+)}{n(\mathrm{H})} \approx 10^{-4}, \ \frac{n(\mathrm{He}^+)}{n(\mathrm{He})} \approx 10^{-14} \quad \text{and} \quad \frac{n(\mathrm{Na}^+)}{n(\mathrm{Na})} \approx 10^3. \tag{2.74}$$

We see that in the solar atmosphere hydrogen is partially ionized, helium is hardly ionized at all, and sodium is predominantly ionized. The degree of ionization is higher in hotter stellar atmospheres. Indeed, if $n_e$ remains at about $10^{19}$ m$^{-3}$, 50% of hydrogen is ionized at about 9000 K, and 50% of helium is ionized at about 15,500 K, as shown in Fig. 2.4.

These considerations help to explain the approximate relation between the surface temperature and the spectral classification of a star:

The spectral classification is based upon the observation of dark lines in the spectrum due to the absorption of photons of particular energies by atoms and ions in the stellar photosphere. For example, the observation of the absorption lines belonging to the Balmer series would imply that the temperature is such that

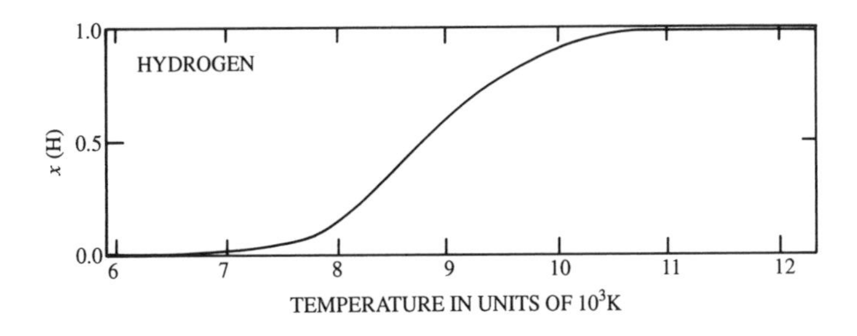


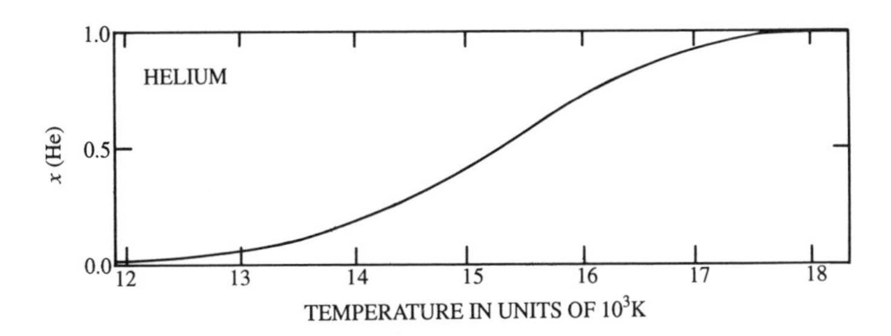


Fig. 2.4 The fractional degree of ionization of hydrogen and of helium as function of the temperature in a gas with a free electron concentration $n_e = 10^{19}\ \mathrm{m}^{-3}$.

hydrogen atoms are present in the $n = 2$ level illustrated in Fig. 2.3; the photo-excitation of these atoms to states with $n = 3$, 4, etc, gives rise to dark lines of the Balmer series. But atoms in the $n = 2$ states will not be present if the temperature is too hot or too cold: if the temperature is too high most of the hydrogen will be ionized and if the temperature is too cold most of the hydrogen will be in the ground state of atomic hydrogen or in the form of molecular hydrogen. As a result, the absorption lines of the Balmer series are only conspicuous in type A and F stars where the surface temperature is between 11,000 K and 6000 K.

Similar considerations apply to the absorption lines of other elements. In brief, the spectrum of hot type O and B stars, with a surface temperature between 30,000 and 12,000 K, are characterized by absorption lines due to the presence of singly ionized helium which do not appear in cooler stars. Absorption lines due to atomic hydrogen are conspicuous in type A and F stars where the surface temperature is between 11,000 and 6000 K. Finally, lines due to the presence of neutral metals are particularly apparent in the spectrum of cooler type G, K and M stars where the temperature is between 6000 and 3000 K.

The spectral lines due to the absorption of photons with particular energies are seen against an opaque and luminous background. This background is produced by the absorption and emission of photons with a continuum of energies in the visible region of the electromagnetic spectrum. For example, electrons in stellar atmospheres emit and absorb photons as they accelerate past ions; these processes, which are usually called bremsstrahlung and inverse bremsstrahlung, are particularly important in hot stellar atmospheres. But cooler atmospheres are rendered opaque and luminous by a process of considerable interest, the continual production and destruction of hydrogen-minus ions.

The $H^-$ ion is a bound state of a proton and two electrons. It is a two-electron system like the helium atom. But the charge on the nucleus is only $Z = 1$ and the second electron is only just bound with a binding energy of only 0.75 eV. Accordingly, photons with an energy as low as 0.75 eV, i.e. a wavelength as long as 1650 nm, are absorbed and emitted by the reactions

$$\gamma + H^- \rightleftharpoons e^- + H. \tag{2.75}$$

However, a gas of hydrogen atoms will absorb or emit visible photons in this way only if free electrons are present. In other words, a transparent gas of hot, un-ionized hydrogen can be made opaque and luminous by the presence of free electrons. We shall now show that a small abundance of easily ionized metallic elements in such a gas can provide these electrons.

We shall model the situation by assuming the presence of metallic elements, denoted by M, which are partially ionized to give a mixture of atoms and pairs of electrons and ions. The concentration of these particles and the fractional ionization of the element $x(M)$ are related by

$$n_e = n(M^+) = x(M)[n(M) + n(M^+)]. \tag{2.76}$$

If we assume that all the metallic elements have the same ionization energy as sodium, the Saha equation gives

$$\frac{[1 - x(M)]}{x(M)^2} \approx \frac{[n(M) + n(M^+)]}{10^{21}T^{3/2}} \exp[60{,}000/T]. \tag{2.77}$$

The dynamic equilibrium concentration of $H^-$ ions is established by the reactions $\gamma + H^- \rightleftharpoons e^- + H$, where the electron concentration is primarily determined by the ionization of metallic atoms. The Saha equation for the ratio of the number of $H^-$ ions to the number of H atoms is

$$\frac{n(H^-)}{n(H)} \approx \frac{n_e}{10^{21}T^{3/2}} \exp[8700/T], \tag{2.78}$$

where the electron concentration $n_e$ is determined by Eqs. (2.76) and (2.77).

A simple numerical calculation, based on Eq. (2.77) and Eq. (2.78), illustrates the coupled roles of easily ionized metals and loosely bound $H^-$ ions in stellar atmospheres. The fraction of metallic atoms ionized $x(M)$ and the concentration of free electrons $n_e$ increase with temperature until nearly all the metallic atoms are ionized at about 4000 K. The concentration of $H^-$ ions reflects this change in the number of free electrons. The number of $H^-$ ions increases with temperature as the electrons become available, but then declines as $n_e$ approaches saturation and as the temperature becomes too hot for the existence of a loosely bound $H^-$ ion. The temperature dependence of $n(H^-)/n(H)$ is illustrated in Fig. 2.5 for the particular case when the abundance of metallic atoms is such that, when they are singly ionized, the concentration of free electrons is $10^{19}$ m$^{-3}$.

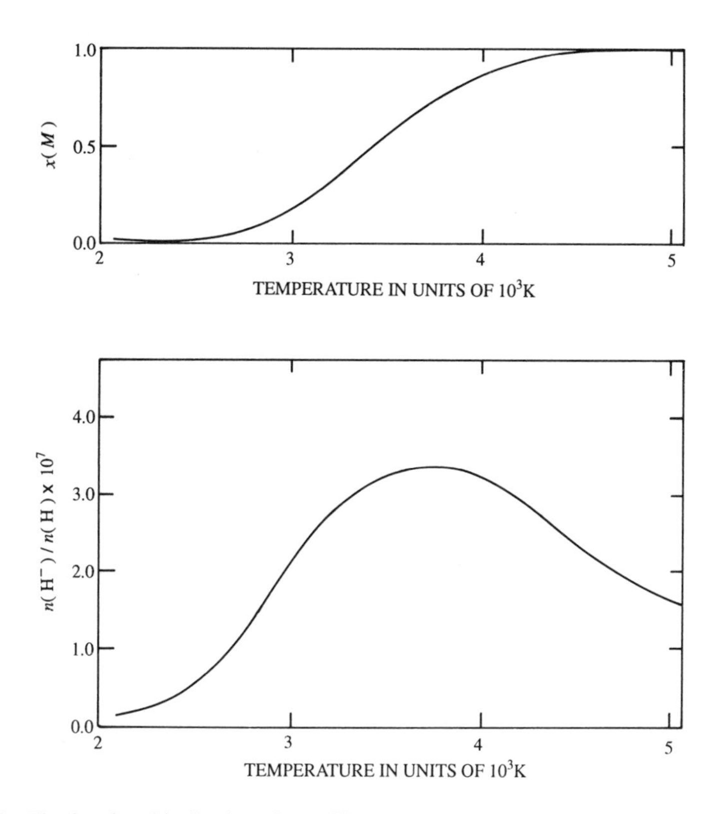


Fig. 2.5 The fractional ionization of metallic atoms like sodium and the ratio of the number of $H^-$ ions to the number of H atoms as a function of temperature. As metallic atoms with an initial concentration of $10^{19}$ m$^{-3}$ are increasingly ionized, the electrons released are captured by hydrogen atoms to form loosely bound $H^-$ ions. These $H^-$ ions dissociate as the temperature increases. The dynamic formation and dissociation of $H^-$ no longer takes place if the temperature falls below 3000 K, and visible radiation is no longer produced and absorbed. Below 3000 K the gas is no longer luminous and opaque.

We note from Fig. 2.5 that at a temperature of about 3000 K there is a small, but significant, abundance of free electrons and $H^-$ ions. At this temperature, $H^-$ ions are continually being produced and destroyed, and as this happens visible radiation is emitted and absorbed. If the temperature is lowered below 3000 K, the abundance of free electrons decreases and the $H^-$ ions in equilibrium with these electrons drops even more precipitously. In particular, $H^-$ ions are no longer being produced and destroyed, and visible radiation is no longer being emitted and absorbed. As a result the gas is no longer luminous and opaque. This phenomenon has an important implication for a stellar atmosphere: The temperature of the observed surface of such an atmosphere is always above 3000 K, or thereabouts.

## 2.6 REACTIONS AT HIGH TEMPERATURE

So far we have focused on the interaction of radiation with atoms. When the temperature becomes comparable with a billion degrees the interaction of radiation with matter gives rise to two new processes: the production of electron–positron pairs and the photodisintegration of atomic nuclei. Both these processes are important in highly evolved massive stars. In this section we indicate how simple and powerful results can be obtained for these processes provided they take place in a system which is close to thermodynamic equilibrium.

### Electron–positron pair production

Electron–positron pairs can be produced when the typical thermal energy, $kT$, is comparable with $m_e c^2$. The equilibrium concentrations of electrons and positrons, $n(e^-)$ and $n(e^+)$, can be found by considering the reactions

$$\gamma + \gamma \rightleftharpoons e^- + e^+, \tag{2.79}$$

noting that the chemical potential of a photon gas is zero, and then equating the sum of the electron and positron chemical potentials to zero,

$$\mu(e^-) + \mu(e^+) = 0. \tag{2.80}$$

If the electrons and positrons form a classical gas, we can use Eq. (2.21) for the chemical potentials to give

$$n(e^-)n(e^+) = 4n_Q^2 \exp[-2m_e c^2/kT], \tag{2.81}$$

where $n_Q$ is the quantum concentration for electrons or positrons.

In a star the concentration of electrons is dominated by the electrons arising from the ionization of the stellar matter. In the hot central regions of an evolved star only traces of hydrogen will remain unburnt, and, according to Eq. (2.69), the electron concentration at density $\rho$ is approximately $n(e^-) \approx \rho/2m_{\rm H}$. As an example, we consider matter with $\rho \approx 10^7$ kg m$^{-3}$ and $T \approx 10^9$ K. Then $n(e^-) \approx 3 \times 10^{33}$ m$^{-3}$ and the equilibrium concentration of positrons given by Eq. (2.81) is $n(e^+) \approx n(e^-)/100$. However, Eq. (2.81) is not valid at higher densities where the electrons are degenerate. In this case pair production is inhibited, because an electron can only be produced if there is an unoccupied quantum state available to be filled. Thus pair production is favoured by high temperature and low density. Accordingly, it is more likely in the centres of very massive stars which attain very high temperatures at comparatively low densities.

The most important astrophysical implication of pair production is that it can lead to the production of neutrinos via

$$\gamma + \gamma \rightleftharpoons e^- + e^+ \rightarrow \nu_e + \overline{\nu}_e. \tag{2.82}$$

Most electron–positron pairs annihilate to yield photons. But about one in $10^{22}$ pairs yield neutrinos. This is one of the mechanisms for neutrino production that can occur in the hot central regions of highly evolved stars. The neutrinos so produced can escape almost unhindered from a stellar interior and thereby provide a very efficient energy loss mechanism.

Energy loss by neutrinos can be important in stars if their core reaches a temperature of $10^9$ K at a density where the electrons are not too degenerate, i.e. less than $10^9$ kg m$^{-3}$, or thereabouts. Note that, even though this energy loss is often called *neutrino cooling*, it does not lead to cooling. Its main effect is to stimulate a faster rate of thermonuclear fusion in order to maintain steady conditions inside the star. As a result, energy loss by neutrinos accelerates the rate of evolution of the star.

**Photodisintegration of nuclei**

A second phenomenon induced by the interaction of radiation at high temperatures is the break-up of atomic nuclei. This process is the analogue of the ionization of atoms which, we recall from Section 2.5, becomes important at about 3000 K. Since nuclear binding energies are typically a million times larger than atomic binding energies, nuclear photodisintegration becomes appreciable at a temperature which is about a million times higher than 3000 K, i.e. when the temperature is about $3 \times 10^9$ K.

We shall see in Section 4.4 that photodisintegration occurs during advanced stages of nuclear burning in massive stars, beginning with neon burning which is initiated by the photodisintegration of $^{20}$Ne via the reaction

$$\gamma + {}^{20}\mathrm{Ne} \rightarrow {}^{16}\mathrm{O} + {}^{4}\mathrm{He}.$$

The $^4$He nuclei released can then be captured by an undissociated $^{20}$Ne nuclei to form $^{24}$Mg. Photodisintegration also plays the key role in silicon burning, the final stage of nuclear burning which leads to the formation of nuclei near iron in the periodic table.

## SUMMARY

### The ideal gas

- In an ideal gas, particles occupy states whose energy is unaffected by the interactions between the particles. The number of quantum states with a momentum of magnitude between $p$ and $p + \mathrm{d}p$ is

$$g(p)\mathrm{d}p = g_s \frac{V}{h^3} 4\pi p^2 \mathrm{d}p. \tag{2.3}$$

The average number of particles in a state with energy $\epsilon_p$ is given by Eq. (2.8) if the particles are identical fermions, and by Eq. (2.9) if the particles are identical bosons; i.e.

$$f(\epsilon_p) = \frac{1}{\exp[(\epsilon_p - \mu)/kT] \pm 1}$$

where the plus sign applies to fermions and the minus sign to bosons.
- The pressure in an ideal gas is given by

$$P = \frac{1}{3V} \int_0^\infty p v_p f(\epsilon_p) g(p) \mathrm{d}p, \tag{2.13}$$

which equals two-thirds the kinetic energy density if the gas particles are non-relativistic and one-third the kinetic energy density if they are ultra-relativistic.
- If the concentration of particles in an ideal gas is low compared with the quantum concentration, their distribution in the quantum states is given by

$$f(\epsilon_p) \approx \exp[-(\epsilon_p - \mu)/kT]. \tag{2.11}$$

These particles form a classical gas. The quantum concentration for non-relativistic particles in such a gas is

$$n_Q = \left[\frac{2\pi mkT}{h^2}\right]^{3/2}, \tag{2.22}$$

and the quantum concentration for ultra-relativistic particles is

$$n_Q = 8\pi \left[\frac{kT}{hc}\right]^3. \tag{2.24}$$

- The chemical potential of a classical ideal gas is

$$\mu = mc^2 - kT \ln\left[\frac{g_s n_Q}{n}\right]. \tag{2.21}$$

**Electrons in stars**

- If the concentration of electrons greatly exceeds the quantum concentration, the electrons form a degenerate gas in which all the electrons fully occupy quantum states with a momentum less than or equal to the Fermi momentum. The Fermi momentum is related to electron concentration by

$$p_F = \left[\frac{3n}{8\pi}\right]^{1/3} h. \tag{2.27}$$

- The equation of state of a non-relativistic, degenerate electron gas is

$$P = K_{NR} n^{5/3}, \quad \text{where} \quad K_{NR} = \frac{h^2}{5m}\left[\frac{3}{8\pi}\right]^{2/3}. \tag{2.31}$$

  This is replaced by

$$P = K_{UR} n^{4/3}, \quad \text{where} \quad K_{UR} = \frac{hc}{4}\left[\frac{3}{8\pi}\right]^{1/3}, \tag{2.34}$$

  if the electrons are predominantly ultra-relativistic.
- The classical, quantum, non-relativistic and ultra-relativistic regimes for an electron gas are illustrated in Fig. 2.2. The properties of electron gases in these different regimes have key roles in stellar evolution.

**Photons in stars**

- Thermal radiation can be considered as a photon gas, a gas of zero mass bosons with zero chemical potential.
- The number of photons per unit volume is

$$n = bT^3 \quad \text{where} \quad b = 2.03 \times 10^7 \text{ K}^{-3} \text{ m}^{-3}. \tag{2.42}$$

- The energy per unit volume in a photon gas is

$$u = aT^4 \quad \text{where} \quad a = 7.565 \times 10^{-16} \text{ J K}^{-4} \text{ m}^{-3}. \tag{2.43}$$

- The radiation pressure is

$$P_r = \frac{1}{3}aT^4. \tag{2.44}$$

This is comparatively small in the sun, but it is more important in more massive stars. According to Eq. (2.49), the ratio of the radiation pressure to the gas pressure in a star of mass $M$ is proportional to $M^2$.

### The Saha equation

- Ionization of atomic hydrogen and recombination,

$$\gamma + \mathrm{H} \rightleftharpoons e^- + \mathrm{H}^+$$

can result in an equilibrium characterized by $\mu(\mathrm{H}) = \mu(\mathrm{e}^-) + \mu(\mathrm{H}^+)$. The equilibrium concentrations are given by

$$\frac{n(\mathrm{H}^+)}{n(\mathrm{H})} \approx \frac{n_{Qe}}{n_e} \exp[-E_i/kT], \tag{2.60}$$

where $E_i$ is the ionization energy of the hydrogen atom.

### Ionization in stars

- The ionization in stellar interiors is almost complete and the number of particles per unit volume in a fully ionized gas is

$$n \approx [1 + 3X_1 + 0.5X_4]\rho/m_{\mathrm{H}}, \tag{2.67}$$

where $X_1$ and $X_4$ are the hydrogen and helium mass fractions. The number of electrons and the number of ions per unit volume are

$$n_e \approx [1 + X_1]\rho/2m_{\mathrm{H}} \quad \text{and} \quad n_i \approx [2X_1 + 0.5X_4]\rho/2m_{\mathrm{H}}. \tag{2.69}$$

- Ionization is partial in stellar atmospheres. Often most of the electrons arise from easily ionized metallic elements. These electrons and the temperature determine the degree of ionization of hydrogen and helium in the atmosphere. Typically 50% of the hydrogen is ionized at 9000 K and 50% of the helium is ionized at 15,500 K; see Fig. 2.4.
- The opaque luminous surface of a star is due to the continual absorption and emission of visible photons. This happens as electrons accelerate past ions, and when loosely bound $\mathrm{H}^-$ ions are formed and broken up. The coupled roles

of easily ionized metallic elements and loosely bound $H^-$ ions indicate that the minimum temperature of the observed surface of a star is about 3000 K; see Fig. 2.5.

## Reactions at high temperature

- Electron–positron pair production becomes significant when $kT$ is comparable with $m_ec^2$ in a non-degenerate electron gas. The equilibrium concentrations of electrons and positrons are given by

$$n(e^-)n(e^+) = 4n_Q^2 \exp[-2m_ec^2/kT]. \tag{2.81}$$

  Pair production leads to neutrino production. Energy loss by neutrinos in massive stars speeds up the rate evolution.
- During the latter stages of the evolution of a massive star, central temperatures above $3 \times 10^9$ K can be reached. At this temperature, and above, high energy thermal photons can break-up atomic nuclei. Photodisintegration of nuclei plays a key role in neon and in silicon burning.

## PROBLEMS 2

2.1 Consider an ideal gas of degenerate, non-relativistic electrons with a concentration $n$ and obtain an expression for the Fermi energy. Assume now that the gas has a temperature $T$ such that the quantum concentration $n_Q$, given by Eq. (2.22), is equal to the actual concentration $n$; quantum effects will be important in such a gas, but the electrons will not be completely degenerate. Find the ratio of $kT$ to the Fermi energy.

2.2 Compare the relative importance of the electrostatic interactions between degenerate electrons and ions in a normal metal with a density of about $10^4$ kg m$^{-3}$ and in a white dwarf with a density of about $10^8$ kg m$^{-3}$. In both cases estimate the temperature below which the electrons are, indeed, degenerate.

2.3 The pressure in an ideal degenerate electron gas is given by Eq. (2.31) if the electrons are non-relativistic, and by Eq. (2.34) if the electrons are predominantly ultra-relativistic. Use the relativistic relation between energy and momentum, $\epsilon_p^2 = p^2c^2 + m^2c^4$, and show that the general expression for the pressure in an ideal degenerate gas is

$$P = K_{UR}n^{4/3}I(x)$$

where $x = p_F/mc$ and

$$I(x) = \frac{3}{2x^4}\left[x(1+x^2)^{1/2}\left(\frac{2x^2}{3} - 1\right) + \ln[x + (1+x^2)^{1/2}]\right].$$

Confirm that, in the appropriate limits, this expression for the pressure reduces to Eq. (2.31) and Eq. (2.34), respectively. (This general expression for the pressure in an ideal degenerate gas will be used in the discussion on white dwarfs in Chapter 6.)

2.4 A stellar atmosphere consists almost entirely of hydrogen. Assume that 50% of the hydrogen molecules are dissociated into atoms and that the pressure is 100 Pa. Given that the binding energy of the hydrogen molecule is 4.48 eV estimate the temperature.

2.5 In the early universe electrons and positrons coexisted with photons at very high temperature. The concentrations of electrons and positrons were approximately equal and were determined by a thermodynamic equilibrium set up by the reactions

$$\gamma + \gamma \rightleftharpoons e^- + e^+.$$

It follows that both the electrons and the positrons formed a gas of ultra-relativistic fermions with zero chemical potential.
Reconsider the calculation for ultra-relativistic bosons with zero chemical potential which led to Eq. (2.42) and Eq. (2.43), and derive the corresponding results for fermions. Show, in particular, that the number of fermions per unit volume and the energy of these fermions are given by

$$n_F = \frac{3}{4}bT^3, \quad \text{and} \quad u_F = \frac{7}{8}aT^4.$$

In fact, the bulk of the pressure in the early universe was due to a gas of photons, electrons, positrons, and three types of neutrinos and antineutrinos. In all, there were 8 types of ultra-relativistic fermions in equilibrium with photons at a high temperature $T$. Bearing in mind that the electrons and positrons have spin half with two polarizations, and neutrinos and antineutrinos have spin half but only one polarization, find an expression for this pressure.

2.6 Consider electron–positron production in a degenerate electron gas with Fermi energy $\epsilon_F$, and derive an expression, analogous to Eq. (2.81), for the equilibrium concentration of positrons. Make a numerical estimate for this concentration in stellar matter at $T = 10^9$ K and $\rho = 10^{10}$ kg m$^{-3}$.

2.7 When the core of a massive star exceeds the Chandrasekhar limit, it collapses. During this collapse, energy is absorbed by the photodisintegration of $^4$He via the reaction

$$\gamma + {}^4\text{He} \rightarrow 2p + 2n.$$

The energy required for this reaction is $Q = 28.30$ MeV. Assume that this reaction is in equilibrium with its inverse, and estimate the temperature at which 50% of the $^4$He is dissociated into nucleons when the density is $10^{12}$ kg m$^{-3}$.

CHAPTER 3

# Heat transfer in stars

There are two basic mechanisms for the transport of heat inside a star. The first mechanism depends upon the random thermal motion of the constituent particles. The particles move, interact and transfer energy from hot to cold regions. If the particles are electrons or ions this process is called thermal conduction. If the particles are photons this process is called radiative diffusion. The second mechanism depends on the collective motion of the constituent particles and is called convective heat transfer. If the temperature gradient is steep enough, heat is transferred from hot to cold regions by rising pockets of hot buoyant fluid and by falling pockets of cool dense fluid. Heat transfer is a complex and difficult subject. We shall focus on the basic ideas, and use these ideas to understand how the heat, generated by nuclear fusion at the centre of a star, is transported to the surface.

## 3.1 HEAT TRANSFER BY RANDOM MOTION

Consider a gas in which the temperature $T$ depends weakly on a coordinate $x$, so that heat flows in the $x$-direction between regions which are approximately in thermodynamic equilibrium. The microscopic mechanism underlying this flow of heat is the random motion of the gas particles. In general, these move with a distribution of speeds, in all possible directions and with a distribution of free paths before they interact. We shall assume, for the sake of simplicity, that one sixth of the particles move in the $x$-direction with a speed $v$ and that they travel a distance

$l$ before they interact. The thermal energy per unit volume at $x$ will be denoted by $u(x)$.

We begin our analysis by considering a surface at a particular value of $x$ and the particles crossing this surface. If there is a temperature gradient, the particles which cross the surface from below will have a different thermal energy from those which cross the surface from above. As a result there is a net transfer of energy across the surface. As indicated in Fig. 3.1, the particles moving from below originate, on average, from a region at $x - l$ and transfer across the surface an energy which is proportional to $u(x - l)$, whereas particles from above originate from $x + l$ and transfer an energy proportional to $u(x + l)$. This implies that the rate of energy transfer across unit area of the surface is given by

$$j(x) \approx \frac{1}{6} v u(x - l) - \frac{1}{6} v u(x + l) \approx -\frac{1}{3} v l \frac{du}{dx}. \tag{3.1}$$

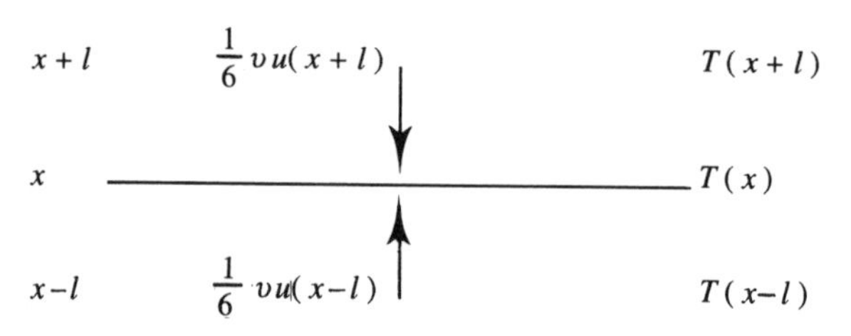


Fig. 3.1 Energy transfer across a surface at $x$ by random thermal motion of particles.

Since $u$ and $T$ are both functions of $x$

$$\frac{du}{dx} = \frac{du}{dT}\frac{dT}{dx} = C\,\frac{dT}{dx}, \tag{3.2}$$

where $C$ is the heat capacity per unit volume. Hence the flux density of heat across the surface at $x$ is directly proportional to the temperature gradient:

$$j(x) = -K\,\frac{dT}{dx} \quad \text{with} \quad K \approx \frac{1}{3} v l C. \tag{3.3}$$

The coefficient $K$ in Eq. (3.3) is the coefficient of thermal conductivity of the gas. A more sophisticated calculation, which takes into account that the particles have a distribution of speeds, directions and free paths, gives a similar result for $K$, but with $v$ and $l$ replaced by the mean speed $\overline{v}$ and the mean free path $\overline{l}$.

We shall first use Eq. (3.3) to describe heat conduction by randomly moving electrons and ions in a plasma. We shall then use it to describe heat conduction by randomly moving photons, a process which is usually called *radiative diffusion*.

**Random motion of electrons and ions**

The importance of thermal conduction by electrons and ions in a plasma can be assessed by using familiar results from the kinetic theory of gases. For classical electrons with concentration $n_e$ at temperature $T$

$$u_e = \frac{3}{2}n_e kT,\ C_e = \frac{3}{2}n_e k \quad \text{and} \quad \overline{v}_e \approx [3kT/m_e]^{1/2}. \tag{3.4}$$

Because electron–electron collisions are less effective at transferring energy than electron–ion collisions, the relevant mean free path in Eq. (3.3) is the mean free path for an electron to collide with an ion. This equals $1/n_i\sigma$, where $n_i$ is the concentration of ions and $\sigma$ is the electron–ion collision cross-section. (The relation between mean free paths and cross-sections is considered in more detail in Section 4.1 of Chapter 4; see Eq. (4.15).) An order of magnitude estimate for the electron–ion cross-section is $\pi r^2$, where $r$ is the distance at which the potential energy of an electron-ion pair is comparable to the thermal kinetic energy; a significant energy transfer to the ion is likely if the electron comes within this distance. For an ion with charge $Z$ this distance is given by

$$\frac{Ze^2}{4\pi\epsilon_0 r} \approx kT. \tag{3.5}$$

Substituting these results into Eq. (3.3) gives the following estimate for the coefficient of thermal conductivity due to electrons

$$K_e \approx \frac{k}{2\pi}\frac{n_e}{n_i}\left[\frac{3kT}{m_e}\right]^{1/2}\left[\frac{4\pi\epsilon_0 kT}{Ze^2}\right]^2. \tag{3.6}$$

The thermal conductivity due to ions, $K_i$, can be obtained from Eq. (3.6) by interchanging $n_e$ and $n_i$, and $m_e$ and $m_i$. If we assume the plasma is fully ionized with $n_e = Zn_i$, we find

$$K_i = \frac{1}{Z^2}\left[\frac{m_e}{m_i}\right]^{1/2} K_e. \tag{3.7}$$

Since $Z > 1$ and $m_i >> m_e$, it follows that $K_i << K_e$, a result which merely reflects the fact that ions are outnumbered by electrons and that they move less quickly than electrons. Thus, the random thermal motion of ions is, in general, a less effective mechanism for heat transfer than the random thermal motion of electrons.

In fact, thermal conductivity by electrons and ions is of minor importance in most stars. White dwarf stars are a notable exception. Here the electrons form a dense, degenerate gas with high thermal conductivity, as in a metal. Equation (3.3) is still applicable, but Eq. (3.4) must be modified to take account of the degeneracy. If the Fermi energy, $\epsilon_F$, is large compared with $kT$, the typical electron speed is increased by a factor of about $(\epsilon_F/kT)^{1/2}$ and the thermal capacity is reduced by a factor of about $kT/\epsilon_F$. The mean free path for an electron collision is also longer in a degenerate gas, because an electron can only be scattered if there is an unoccupied state available to be filled. The net result is that heat in the interior of a white dwarf is conducted very efficiently by degenerate electrons. Indeed, to a first approximation, a white dwarf has an interior of high conductivity, at a temperature which is almost uniform, surrounded by an insulating jacket of non-degenerate electrons and ions. The heat transfer through this jacket will be considered in Section 3.4.

### Random motion of photons

We can also use Eq. (3.3) to assess the importance of radiative diffusion, the thermal conduction of heat by photons. We recall from Section 2.3 that thermal photons move with the speed of light and, according to Eq. (2.43), form a gas with an energy density and a thermal capacity given by

$$u_r = aT^4 \quad \text{and} \quad C_r = 4aT^3. \tag{3.8}$$

Hence the heat flux density due to radiative diffusion is

$$j(x) = -K_r \frac{dT}{dx} \quad \text{with} \quad K_r \approx \frac{4}{3} c\bar{l}aT^3, \tag{3.9}$$

where $K_r$ can be thought of as the coefficient of thermal conduction due to the random motion of photons.

To proceed further we need to know $\bar{l}$, the mean free path for a photon collision in stellar matter. The simplest situation occurs at the high temperatures and the comparatively low densities found in the interiors of massive main sequence stars. Here the dominant process is Thomson scattering by electrons, in which case

$$\bar{l} = \frac{1}{n_e \sigma_T} \quad \text{where} \quad \sigma_T = \frac{8\pi}{3}\left[\frac{e^2}{4\pi\epsilon_o m_e c^2}\right]^2. \tag{3.10}$$

The Thomson scattering cross-section, $\sigma_T$, can be derived by considering the classical radiation of an accelerating electron, or more generally from quantum electrodynamics. The coefficient for thermal conduction by photons can be found by substituting the mean free path into Eq. (3.9). It is instructive to compare the result with Eq. (3.6), the corresponding coefficient for conduction by electrons. Straightforward algebra gives

$$\frac{K_r}{K_e} \approx \sqrt{3}Z\frac{P_r}{P_e}\left[\frac{m_e c^2}{kT}\right]^{5/2}, \tag{3.11}$$

where $P_r$ and $P_e$ are the radiation and electron pressures, given by Eqs. (2.44) and (2.19), and we have again assumed that $n_e = Zn_i$. To illustrate we consider conditions typical to the solar interior: for a hydrogen plasma at $6 \times 10^6$ K and $1.4 \times 10^3$ kg m$^{-3}$,

$$kT \approx 10^{-3}\, m_e c^2,\ P_r = 3 \times 10^{11}\ \text{Pa} \quad \text{and} \quad P_e = 7 \times 10^{13}\ \text{Pa}.$$

Substitution into Eq. (3.11) gives $K_r \approx 2 \times 10^5\, K_e$. We conclude that radiative diffusion is a more effective mechanism for heat transfer in the sun than thermal conduction by electrons.

This conclusion still holds when we take into account the absorption of photons in the sun. Conservation of energy and momentum implies that a photon cannot be absorbed by an interaction with a free particle. In practice, photon absorption normally involves an interaction with an electron in the presence of an ion, and as such it becomes increasingly important at higher density and lower temperature. If the interacting electron is initially bound to the ion we have a process called *bound–free absorption*, and if the electron is initially unbound we have *free–free absorption*. These processes are also called photo-ionization and inverse-bremsstrahlung.

Both free–free and bound–free absorption lead to a mean free path which varies with the frequency of the photon. Accordingly the analysis leading to Eqs. (3.3) and (3.9) must be modified. We recall that Eq. (2.46) describes the black body radiation due to photons with a frequency between $\nu$ and $\nu + d\nu$. The energy density and the thermal capacity due to the photons in this frequency range are given by

$$u_\nu d\nu = \frac{h\nu}{\exp(h\nu/kT) - 1} 8\pi \frac{\nu^2}{c^3} d\nu \quad \text{and} \quad C_\nu d\nu = \frac{\partial u_\nu}{\partial T} d\nu. \tag{3.12}$$

If $\bar{l}_\nu$ is the mean free path at frequency $\nu$, the coefficient of conduction due to photons of all frequencies is

$$K_r = \int_0^\infty \frac{1}{3} c \bar{l}_\nu C_\nu d\nu. \tag{3.13}$$

We conclude that Eq. (3.9) can still be used to describe radiative diffusion provided the mean free path is averaged over frequency as follows:

$$\bar{l} = \frac{\int_0^\infty \bar{l}_\nu C_\nu d\nu}{4aT^3}. \tag{3.14}$$

This average, which is called the Rosseland average, is likely to be dominated by contributions at frequencies near $2.8kT/h$ where $C_\nu$ is a maximum, and at frequencies where $\bar{l}_\nu$ is large; i.e. where the stellar material is almost transparent.

Regardless of the dominant mechanism for photon scattering or absorption, the photon mean free path is determined by the probability of an interaction with either an electron or an ion. This depends upon the concentration of electrons and ions, $n_e$ and $n_i$, and interaction cross-sections, $\sigma_e$ and $\sigma_i$. The probability of interaction in a distance d$x$ is equal to $(n_e\sigma_e + n_i\sigma_i)$d$x$, and the mean free path is

$$\bar{l} = \frac{1}{n_e\sigma_e + n_i\sigma_i}. \tag{3.15}$$

Since $n_e$ and $n_i$ are both proportional to the mass density of the stellar material, $\rho$, it is customary to write $\bar{l} = 1/\rho\kappa$ and specify the radiative transfer properties in terms of $\kappa$, the opacity of the material. In particular, the flux density of radiant heat, given by Eq. (3.9), is rewritten as

$$j(x) = -\frac{4ac}{3}\frac{T^3}{\rho\kappa}\frac{\mathrm{d}T}{\mathrm{d}x}. \tag{3.16}$$

We shall not consider in detail the complex mechanisms underlying the opacity of stellar material. We shall merely indicate the most important features:

Bound–free absorption is important at low temperatures where a large fraction of the atoms are only partially ionized. Free–free absorption dominates at higher temperatures where ionization nears completion. These mechanisms give a frequency-averaged opacity which increases with density and decreases with temperature roughly in accordance with

$$\kappa \propto \rho\, T^{-3.5}. \tag{3.17}$$

This is known as Kramers' law. Electron scattering provides a constant background opacity which becomes predominant at high temperatures and low densities. This constant opacity can be found using Eqs. (3.10) and (2.69),

$$\kappa_{es} = n_e\sigma_T/\rho = (1 + X_1)\,\sigma_T/2m_{\mathrm{H}} \approx (1 + X_1)\,0.02\ \mathrm{m}^2\ \mathrm{kg}^{-1} \tag{3.18}$$

where $X_1$ is the mass fraction of hydrogen in the stellar material.

By way of numerical illustration, the opacity of solar material at a density of $10^4$ kg m$^{-3}$ and a temperature of $2 \times 10^6$ K, is about 10 m$^2$ kg$^{-1}$, corresponding to a photon mean free path, $1/\kappa\rho$, of about $10^{-5}$ m. At a higher temperature of $1 \times 10^7$ K the opacity is smaller and the mean free path longer, about 0.1 m$^2$ kg$^{-1}$ and $10^{-3}$ m.

## 3.2 HEAT TRANSFER BY CONVECTION

In the last section we considered how the random motion of photons, electrons and ions in a material leads to the conduction of heat. However in the presence of a force field, heat may be transferred by the collective motion of the constituent particles. Gravity provides this force field in a star. A rising pocket of stellar gas may sometimes find itself in a cooler and more dense environment, and it will continue to rise because of its buoyancy. A falling pocket of gas will continue to fall if it finds itself in a warmer less dense environment. Complex and unpredictable currents can be set up which convect heat very efficiently from hot to cold regions. Indeed, convection is so efficient that it will dominate other heat transfer mechanisms. However, convection only takes place if the magnitude of temperature gradient exceeds a certain critical value.

### Critical condition for convection

Consider an ideal gas in a gravitational field. We shall denote the temperature, pressure and density by $T$, $P$ and $\rho$ at height $x$, and by $T+\Delta T$, $P+\Delta P$ and $\rho+\Delta\rho$ at height $x+\Delta x$. Because the gas satisfies the ideal gas law, we have $\rho \propto P/T$ and

$$\frac{\Delta\rho}{\rho} = \frac{\Delta P}{P} - \frac{\Delta T}{T}. \tag{3.19}$$

HEIGHT $x + \Delta x$ — $T + \delta T$, $P + \delta P$, $\rho + \delta\rho$ — $T + \Delta T$, $P + \Delta P$, $\rho + \Delta\rho$

HEIGHT $x$ — $T$, $P$, $\rho$ — $T$, $P$, $\rho$

Fig. 3.2 Displacement of pocket of gas from height $x$ to $x + \Delta x$.

Now consider a pocket of gas at height $x$, as shown in Fig. 3.2. We shall assume that the temperature, pressure and density of the gas in this pocket and of the surrounding gas are matched. In general they will not match if the pocket is displaced to a height $x + \Delta x$. We shall denote the changes in the temperature, pressure and density of the displaced pocket by $T + \delta T$, $P + \delta P$ and $\rho + \delta\rho$. It is reasonable to assume that the pressure inside the pocket responds rapidly to the new environment so that $\delta P = \Delta P$. We shall also assume that there is insufficient time for heat conduction to the environment and that the displaced pocket of gas expands adiabatically until its pressure matches the surrounding pressure. For an adiabatic process $P \propto \rho^{\gamma}$, so that

$$\frac{\delta\rho}{\rho} = \frac{1}{\gamma}\frac{\delta P}{P}. \tag{3.20}$$

The pocket will be buoyant, and will continue to rise, if it contains gas which is less dense than the surrounding gas; i.e. convection is possible if

$$\delta\rho < \Delta\rho, \quad \text{or} \quad \frac{1}{\gamma}\frac{\delta P}{P} < \frac{\Delta P}{P} - \frac{\Delta T}{T}. \tag{3.21}$$

We can set $\delta P = \Delta P$, because the pressure within the pocket responds quickly to match the surroundings, and rewrite the condition for convection as

$$\frac{\Delta T}{T} < \frac{(\gamma - 1)}{\gamma}\frac{\Delta P}{P}. \tag{3.22}$$

In other words, the critical temperature gradient for convection is given by

$$\frac{\mathrm{d}T}{\mathrm{d}x} < \frac{(\gamma - 1)}{\gamma}\frac{T}{P}\frac{\mathrm{d}P}{\mathrm{d}x}. \tag{3.23}$$

Note that the temperature and the pressure gradients are both negative in this equation. Convection requires the temperature to fall off rapidly with height. This fall-off is determined by the value of the adiabatic index $\gamma$ and the fall-off in the pressure.

The reader may recall that the adiabatic index of an ideal classical gas is related to the number of classical degrees of freedom of the constituent gas particles. In particular, if there are $s$ classical degrees of freedom, each with an average thermal energy of $\frac{1}{2}kT$, then

$$\gamma = \frac{C_P}{C_V} = \frac{1 + s/2}{s/2}. \tag{3.24}$$

For gas particles with just three translational degrees of freedom we have $s = 3$ and hence $\gamma = 5/3$. But $\gamma$ is smaller if the number of degrees of freedom is larger; in fact $\gamma$ approaches 1 as $s$ becomes large. Thus, if the gas particles can absorb heat by exciting internal degrees of freedom such as rotation or vibration, $\gamma$ is smaller and the critical temperature gradient for convection, Eq. (3.23), becomes less steep. This is also the case if heat can be absorbed by the dissociation of molecules or by the ionization of atoms.

The other important factor in Eq. (3.23), the fall-off in pressure with height, depends on the strength of gravity. In particular, if we assume hydrostatic equilibrium in a region where the acceleration due to gravity is $g$,

$$\frac{\mathrm{d}P}{\mathrm{d}x} = -g\rho(x). \tag{3.25}$$

We note that in regions where $g$ is small the pressure falls off gradually and convection is more easily induced.

In practice, convection currents transfer heat very effectively. Indeed, the process is so efficient that, in many circumstances, all the heat generated can be transported a soon as the temperature gradient reaches the critical value given by Eq. (3.23).

## 3.3 TEMPERATURE GRADIENTS IN STARS

The temperature gradient at a point inside a star is determined by the rate of flow of energy towards the surface and the mechanism governing this energy flow. In practice, the most important mechanisms for the flow of energy in stars are often radiative diffusion and convection.

Let $L(r)$ denote the rate at which energy flows outwards through a spherical surface of radius $r$ within the star. The release of nuclear energy in the hot, centre of the star implies that $L(r)$ increases with $r$ until a region is reached in which no energy is being released. Indeed, if $\varepsilon(r)$ denotes the nuclear power generated per unit volume at $r$, then the power produced in a shell bounded by $r$ and $r + \mathrm{d}r$ is $\varepsilon(r)4\pi r^2\mathrm{d}r$. Because this is added to the outward power flow,

$$\frac{\mathrm{d}L}{\mathrm{d}r} = 4\pi r^2 \varepsilon(r). \tag{3.26}$$

Outside any central generating regions, $L(r)$ becomes constant and approaches the surface luminosity of the star.

We begin by assuming that radiative diffusion is the dominant heat transfer mechanism. In this case the total outward power flow is $L(r) = 4\pi r^2\, j(r)$, with $j(r)$ given by Eq. (3.16). Hence

$$\frac{L(r)}{4\pi r^2} = -\frac{4ac[T(r)]^3}{3\rho(r)\kappa(r)}\frac{\mathrm{d}T}{\mathrm{d}r}, \tag{3.27}$$

where the temperature $T$, the density $\rho$ and the opacity $\kappa$ depend on $r$. In fact, it is more useful to think in terms of how the star manages to transport the power generated in the interior towards the surface. If it does so by radiative diffusion, it sets up a temperature gradient given by

$$\left[\frac{\mathrm{d}T}{\mathrm{d}r}\right]_{rad} = -\frac{3\rho(r)\kappa(r)}{4ac[T(r)]^3}\frac{L(r)}{4\pi r^2}. \tag{3.28}$$

The implications of Eq. (3.28) for heat transfer in the sun can be assessed by inserting appropriate numerical values. We can assume that the power flow in the sun reaches a constant value equal to the surface luminosity of $4 \times 10^{26}$ W at a distance of about $0.4R_\odot$ from the centre. If we use the following estimates for the temperature, density and opacity at this distance,

$$T \approx 5 \times 10^6 \text{ K}, \quad \rho \approx 5 \times 10^3 \text{ kg m}^{-3} \quad \text{and} \quad \kappa \approx 0.5 \text{ m}^2 \text{ kg}^{-1},$$

we obtain a temperature gradient of about $-0.03$ K m$^{-1}$. We note that the fractional change in temperature over a distance comparable with the photon mean free path, which in this case is 0.4 mm, is only $2 \times 10^{-12}$. This indicates that the basic approximation underlying radiative diffusion is valid: The solar interior is dense and opaque, and radiation can indeed diffuse slowly to and from regions which are in local thermodynamic equilibrium.

However, radiative diffusion will not be the dominant mechanism for heat transfer if the temperature gradient reaches the critical value for the onset of convection. According to Eq. (3.23) this critical temperature gradient is

$$\left[\frac{dT}{dr}\right]_{conv} = \frac{(\gamma - 1)}{\gamma} \frac{T}{P} \frac{dP}{dr}, \tag{3.29}$$

where the pressure gradient is determined by hydrostatic equilibrium,

$$\frac{dP}{dr} = -\frac{Gm(r)\rho(r)}{r^2}. \tag{3.30}$$

In practice, convection dominates radiative diffusion whenever the temperature gradient reaches the critical value given by Eq. (3.29). Indeed, convection is so efficient that almost any amount of power can be transported and the temperature gradient seldom needs to be steeper than this critical value. Convection is particularly important in ionization zones and in the cores of massive main sequence stars.

Ionization zones occur in the surface layers of stars, where atoms and ions are continuously absorbing and releasing energy by ionization and recombination. Convection is favoured for two reasons. First the opacity $\kappa$ will be large and the temperature gradient for radiative transfer, Eq. (3.28), is steep. Second the temperature gradient needed for the onset of convection, Eq. (3.29), is not steep because the adiabatic index $\gamma$ is close to one; in more physical terms, convection is favoured because a rising pocket of gas does not cool so much and is more likely to remain buoyant if electron recombination can provide some of the energy needed to expand the gas.

There is a convection zone in the sun located just below the photosphere. The thickness is about 0.1 to $0.2R_{\odot}$. Here hot pockets of partially ionized gas rise and cooler pockets of gas sink back down. As a result there appears at the base of the photosphere bright, irregular and transient formations, called granules. The convected energy is dissipated in the photosphere and then transferred to the solar surface by radiative diffusion.

Convection can also be important in the central energy generating regions of stars. The most favoured situation occurs when thermonuclear power is generated in a small region near the centre. In this case, large amounts of energy flow through a region where the acceleration due to gravity is low; the pressure falls off gradually, and a rising pocket of gas is more likely to remain buoyant because it need not expand much.

We can be more quantitative by focusing on $L(r)/m(r)$, the power that is generated per unit mass within a core of radius $r$. If this exceeds a critical value, the core will become convective. To find this critical value we set the radiative temperature gradient, given by Eq. (3.28), equal to the critical gradient for convection, Eq. (3.29) and use Eq. (3.30) to give

$$\frac{3\rho\kappa}{4acT^3}\frac{L(r)}{4\pi r^2} = \frac{(\gamma-1)}{\gamma}\frac{T}{P}\frac{Gm(r)\rho}{r^2}.$$

If we tidy up by replacing $aT^4/3$ by the radiation pressure, $P_r$, we find that the value of $L(r)/m(r)$ needed for convection is

$$\left[\frac{L(r)}{m(r)}\right]_{crit} = \frac{(\gamma-1)}{\gamma}\frac{16\pi Gc}{\kappa}\frac{P_r}{P}. \tag{3.31}$$

If $L(r)/m(r)$ is below this value, energy can be transported from the core by radiative diffusion without inducing convection. If it exceeds this value convection dominates. Thus, a convective core of radius $r$ and mass $m(r)$ is produced if the power generated per unit mass within $r$ exceeds the limit set by Eq. (3.31).

Convection occurs in the cores of massive main sequence stars, where hydrogen burning takes place by the carbon–nitrogen cycle. This process, which will be considered in Chapter 4, is very temperature dependent; in fact, the power generated is proportional to $T^{17}$. As the temperature falls off with $r$ near the centre of the star, nuclear power generation falls off extremely rapidly to give a small generating region in which convection dominates. The central generating region of less massive stars, like the sun, are larger and convection is less likely. This is because hydrogen burning in such stars is via the proton–proton chain which is less temperature dependent than the carbon–nitrogen cycle.

Most models of the sun indicate that convection is not important in the solar core. For example, if we evaluate the right-hand side of Eq. (3.31) using values appropriate to the solar core, $\gamma = 5/3$, $P = 1.73 \times 10^{16}$ Pa, $T = 13.7 \times 10^6$ K and $\kappa = 0.138$ m$^2$ kg$^{-1}$, we find that convection occurs only if the central power generation per unit mass is greater than $1.5 \times 10^{-3}$ W kg$^{-1}$. In practice, the power generated is expected to be about $1.35 \times 10^{-3}$ W kg$^{-1}$, just less than the critical value for convection.

## 3.4 COOLING OF WHITE DWARFS

We shall end this chapter by considering the physics underlying the steady decline in the temperature and luminosity of a white dwarf. Our primary purpose is to illustrate the role of heat transfer in stars. Our secondary purpose is to indicate how the age of a white dwarf can be estimated from its luminosity. This is possible because a white dwarf is an inert star, a dead body with no internal power source. The time of death, as every detective knows, can be deduced from the temperature of the corpse. This time can be used to estimate the age of the white dwarf and also the age of the star system to which the white dwarf belongs.

For the most part, a white dwarf is composed of a dense system of classical ions and degenerate electrons, surrounded by a thin envelope of classical gas particles. The star cools predominantly by the conduction of heat by electrons in the interior, and by the diffusion of radiation through the outer envelope. The cooling time is long because of the high thermal energy of the ions in the interior and the high opacity of the gas composing the envelope. In fact, the time scale for cooling is about a billion years, long enough to ensure that many white dwarfs have not yet faded from view, but short enough to ensure that most white dwarfs have low luminosities.

We shall consider a simple model for a cooling white dwarf consisting of a hot, metal-like, sphere surrounded by an insulating jacket of ionized gas. We shall assume that the temperature of the interior is almost uniform because of the high thermal conductivity of degenerate electrons; such electrons transfer energy over long free paths because they can only be scattered into unoccupied quantum states. The temperature of this isothermal interior will be denoted by $T_I$. We shall also assume that the thermal energy of the ions, typically $\frac{3}{2}kT_I$ per ion, is lost as heat is transported across the outer envelope, mostly by radiative diffusion. Hence, the insulating properties of the outer envelope control the energy loss to outer space and thereby determine the relation between the luminosity $L$ of the star and the steadily declining internal temperature $T_I$. We note that, as energy is lost, there is little change in the structure of the star because it is supported by degenerate electrons which cannot lose energy.

Our first task is to consider the variation of the pressure, temperature and density in the outer envelope of the white dwarf. We assume that the ionized gas in the envelope is classical and ideal with an equation of state $P = \rho kT/\overline{m}$. There is a pressure gradient determined by hydrostatic equilibrium, Eq. (3.30), and a temperature gradient produced by the flow of heat towards the surface, which we assume is governed by radiative diffusion, Eq. (3.28). Hence

$$\frac{\mathrm{d}P}{\mathrm{d}r} = -\frac{GM\rho(r)}{r^2} \quad \text{and} \quad \frac{\mathrm{d}T}{\mathrm{d}r} = -\frac{3\rho(r)\kappa(r)}{4ac[T(r)]^3}\frac{L}{4\pi r^2}. \tag{3.32}$$

Because there is no energy generation, $L$ is the surface luminosity. Also, $m(r)$ has been replaced by $M$, the total mass of the star, because most of the mass is concentrated within the envelope. These two equations can be combined to give

$$\frac{dP}{dT} = \left[\frac{16\pi acG}{3}\frac{M}{L}\right]\frac{T^3}{\kappa}. \tag{3.33}$$

The opacity of the ionized gas in the outer envelope depends on the temperature, density and chemical composition. In this calculation, we shall assume that the opacity is due to bound-free absorption and that 90% of the mass is helium and that 10% is in the form of heavier elements. An appropriate opacity is then given by

$$\kappa = \kappa_0 \rho T^{-3.5} = 4.34 \times 10^{19} \rho T^{-3.5} \text{ m}^2 \text{ kg}^{-1}, \tag{3.34}$$

an expression consistent with Kramers' law, Eq. (3.17). We can use the ideal gas equation to rewrite the opacity in terms of the temperature and pressure

$$\kappa = \left[\frac{\kappa_0 \overline{m}}{k}\right] P T^{-4.5}. \tag{3.35}$$

Substitution into Eq. (3.33) gives the following differential equation relating the pressure and temperature in the envelope:

$$\frac{dP}{dT} = C\,\frac{T^{7.5}}{P}, \quad \text{with} \quad C = \left[\frac{16\pi acGk}{3\kappa_0 \overline{m}}\frac{M}{L}\right]. \tag{3.36}$$

If we integrate and use the boundary condition that $P = 0$ when $T = 0$, we find that

$$\frac{P^2}{2} = C\,\frac{T^{8.5}}{8.5}. \tag{3.37}$$

The pressure, temperature and density increase as we go deeper into the white dwarf. We are particularly interested in the density of the electrons, because these particles will become degenerate as the interior of the white dwarf is approached. This density can be found by noting that two-thirds of the particles in the ionized gas of the envelope are electrons. Thus, electrons provide two-thirds of the pressure with a number density given by

$$n_e = \frac{2}{3}\frac{P}{kT}. \tag{3.38}$$

If this is combined with Eq. (3.37), we find

$$n_e = \frac{2}{3k}\left[\frac{C}{4.25}\right]^{1/2} T^{13/4}. \tag{3.39}$$

The electrons will no longer form a classical gas when $n_e$ approaches the quantum concentration given by Eq. (2.22); i.e. when $n_e$ approaches

$$n_Q = \left[\frac{2\pi m_e kT}{h^2}\right]^{3/2}. \tag{3.40}$$

The transition between the classical electron gas in the envelope and the quantum electron gas in the interior occurs when $n_e \approx n_Q$. The highly conducting, isothermal interior is reached when the electrons become degenerate with $n_e >> n_Q$. In particular, we can obtain an approximate expression for the temperature of the isothermal interior, $T_I$, by assuming that at this temperature $n_e = 10n_Q$. Then Eqs. (3.39) and (3.40) imply that

$$10\left[\frac{2\pi m_e kT_I}{h^2}\right]^{3/2} = \frac{2}{3k}\left[\frac{C}{4.25}\right]^{1/2} T_I^{13/4}. \tag{3.41}$$

If we use the definition of the constant $C$ given in Eq. (3.36), and if we use the sun as a standard of mass and luminosity, we find the following estimate for $T_I$:

$$T_I \approx (7 \times 10^7 \text{ K})\left[\frac{L/L_\odot}{M/M_\odot}\right]^{2/7}. \tag{3.42}$$

Finally, we may rearrange Eq. (3.42) and express the luminosity of a white dwarf in terms of its mass and the temperature of its isothermal interior,

$$L \approx \left[\frac{T_I}{7 \times 10^7 \text{ K}}\right]^{7/2}\left[\frac{M}{M_\odot}\right] L_\odot. \tag{3.43}$$

We note that this approximate relation between the luminosity $L$ and the internal temperature $T_I$ arises because the insulating envelope of the white dwarf controls the loss of energy into outer space.

As already mentioned the energy source for the luminosity of a white dwarf is the thermal energy of the classical ions in the interior. This energy store is very large. For example, if a white dwarf of mass $M$ contained carbon ions in the form of a classical gas, the thermal energy would be

$$E \approx \frac{3}{2}NkT_I = \frac{3}{2}\left[\frac{M}{12m_\text{H}}\right]kT_I. \tag{3.44}$$

This equals $8 \times 10^{40}$ J for a star of mass $0.4M_\odot$ at $10^8$ K. In practice, as the white dwarf cools, the ions crystallize to form a lattice. The specific heat will increase from $\frac{3}{2}Nk$ to $3Nk$, and then decrease as the temperature falls below the Debye temperature of the solid.

Given the luminosity of a white dwarf, Eq. (3.43), and its internal energy store, Eq. (3.44), we can find its cooling rate. If we equate the rate of decrease in the internal energy to the luminosity we find that

$$\frac{dT_I}{dt} = -\alpha \left[ \frac{T_I}{7 \times 10^7 \text{ K}} \right]^{7/2} \quad \text{with} \quad \alpha \approx \frac{2}{3k} \left[ \frac{12m_{\rm H}}{M_\odot} \right] L_\odot \approx 6 \text{ K per year.} \quad (3.45)$$

This simple differential equation may be integrated to give an expression for the internal temperature of a white dwarf as a function of time. This expression may then be substituted into Eq. (3.43) to give the luminosity as a function of time. The initial temperature and luminosity are determined by the events which led up to the formation of the white dwarf. For example, if the white dwarf was formed following the completion of helium burning, the internal temperature will be about $10^8$ K, and the initial luminosity will be about $L_\odot$ if the mass is $0.4M_\odot$.

Figure 3.3 illustrates the declining luminosity of a carbon white dwarf of mass $0.4M_\odot$ with an initial internal temperature of $10^8$ K. Note that the calculated time-scale for cooling from a luminosity of about $L_\odot$ to $10^{-4}L_\odot$ is a billion years.

This elementary calculation only gives a rough guide to the cooling of white dwarfs. Detailed comparison between theory and observation requires a careful analysis of the thermal properties of the ions, heat loss by neutrino emission and energy release by sedimentation under gravity; see Shapiro and Teukolsky (1983). As mentioned earlier, such a comparison is of practical use in astronomy as a way of estimating the age of white dwarfs and the age of the star systems to which they belong.

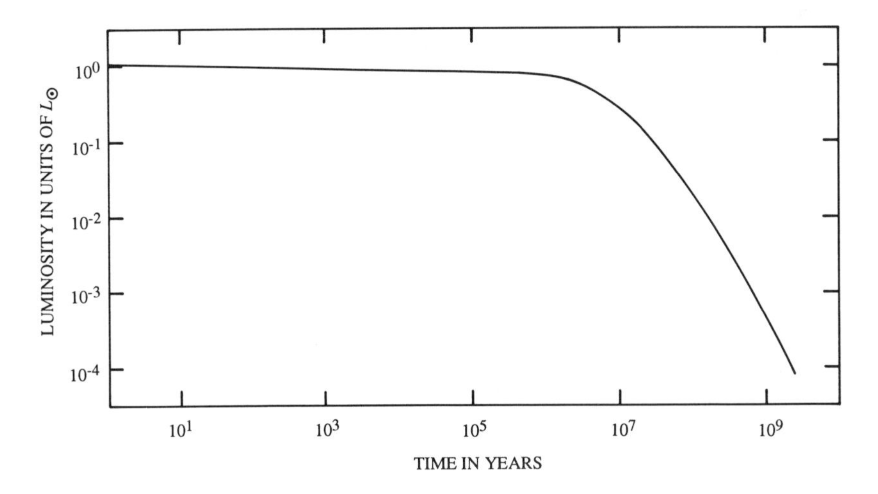


Fig. 3.3 The declining luminosity of a cooling carbon white dwarf of mass $0.4M_\odot$ with an initial internal temperature of $10^8$ K.

## SUMMARY

### Heat transfer by random motion

- The flux density of heat due to the random thermal motion of particles with mean speed $\bar{v}$, mean free path $\bar{l}$, and heat capacity per unit volume $C$ is given by

$$j(x) = -K\frac{dT}{dx} \quad \text{with} \quad K \approx \frac{1}{3}\bar{v}\bar{l}C. \tag{3.3}$$

  Random motion by electrons, ions and photons can lead to the conduction of heat.
- Conduction by photons, or radiative diffusion, is especially important. It leads to a radiative flux density given by

$$j(x) = -K_r\frac{dT}{dx} \quad \text{with} \quad K_r \approx \frac{4}{3}c\bar{l}aT^3. \tag{3.9}$$

- Transfer of heat by radiative diffusion is usually described in terms of the opacity, $\kappa = 1/\rho\bar{l}$. In particular, the radiant flux density is given by

$$j(x) = -\frac{4ac}{3}\frac{T^3}{\rho\kappa}\frac{dT}{dx}. \tag{3.16}$$

  The constant background opacity due to electron scattering is given by

$$\kappa_{es} \approx (1 + X_1)\, 0.02 \text{ m}^2 \text{ kg}^{-1}. \tag{3.18}$$

  At high density and low temperature, bound–free and free–free absorption give rise to an opacity which has a density and temperature dependence given by Kramers' law,

$$\kappa \propto \rho\, T^{-3.5}. \tag{3.17}$$

### Heat transfer by convection

- A rising pocket of gas will remain buoyant and continue to rise, and a falling pocket of gas will continue to fall, if the temperature gradient in a classical ideal gas is given by

$$\frac{dT}{dx} < \frac{(\gamma - 1)}{\gamma}\frac{T}{P}\frac{dP}{dx}. \tag{3.23}$$

  Convection is usually the dominant heat transfer mechanism once this critical temperature gradient is reached.

### Temperature gradients in stars

- If the outward power flow in a star is governed by radiative diffusion, the temperature gradient is given by

$$\left[\frac{dT}{dr}\right]_{rad} = -\frac{3\rho(r)\kappa(r)}{4ac[T(r)]^3}\frac{L(r)}{4\pi r^2}. \tag{3.28}$$

- If the temperature gradient reaches the critical value given by

$$\left[\frac{dT}{dr}\right]_{con} = \frac{(\gamma-1)}{\gamma}\frac{T}{P}\frac{dP}{dr}, \tag{3.29}$$

convection is the dominant mechanism for heat transfer.

### Cooling of white dwarfs

- The important problem of the cooling of white dwarfs illustrates many of the ideas introduced in this chapter. It involves heat transfer by radiative diffusion through the outer layers of the star, and heat conduction by degenerate electrons through the interior of the star.

## PROBLEMS 3

3.1 Show that, if the frequency and temperature dependence of the mean free path for a photon is given by

$$\bar{l}_\nu \propto \nu^3\, T^{1/2},$$

then the frequency averaged opacity satisfies Kramers' law, Eq. (3.17).

3.2 The opacity depends on the chemical composition of the stellar material. Explain why the free–free opacity is proportional to $(X_1 + X_4)\,(1 + X_1)$, and the bound–free opacity is proportional to $X_A(1 + X_1)$, where $X_1, X_4$ and $X_A$ are the mass fractions of hydrogen, helium and heavier elements.

3.3 Show that heat transfer by radiative diffusion implies a non-zero gradient for the radiation pressure which is proportional to the radiant heat flow. Bearing in mind that the magnitude of the force per unit volume in a fluid due to the pressure is equal to the pressure gradient, find the radiant heat flux density which can, by itself, support the atmosphere of a star with surface gravity $g$. Hence show that a star of mass $M$ has maximum luminosity given by

$$L_{max} = 4\pi cGM/\kappa,$$

where $\kappa$ is the opacity near the surface. Obtain a numerical estimate for this luminosity by assuming that the surface is hot enough for the opacity to be dominated by electron scattering. (This maximum luminosity is called the Eddington luminosity.)

3.4 Recall that the adiabatic index, $\gamma$, is the ratio of the heat capacities at constant pressure and at constant volume. Show that, for an ideal classical gas, the critical temperature for the onset of convection, Eq. (3.23), can be written as

$$\left[\frac{\mathrm{d}T}{\mathrm{d}x}\right]_{conv} = -\frac{g}{C_P},$$

where $C_P$ is the thermal capacity per unit mass at constant pressure and $g$ is the acceleration due to gravity. (Note that if the thermal capacity is high because of the absorption of heat by the excitation and/or the dissociation of the constituent particles, then the temperature gradient needed for convection is less steep.)

3.5 The approximate temperature and pressure profiles in the outer envelope of a white dwarf were found in Section 3.4 by assuming hydrostatic equilibrium and heat flow by radiative diffusion. Show that the results obtained justify the neglect of convection.

3.6 Use Eq. (3.37) and show that the radiative temperature gradient in the outer envelope of classical gas surrounding a white dwarf is given by

$$\frac{\mathrm{d}T}{\mathrm{d}r} = -\frac{GM\overline{m}}{4.25r^2k}.$$

Consider a white dwarf with mass $M = 0.4\,M_\odot$ and radius $R = R_\odot/100$ with an internal temperature of $10^7$ K, and estimate the thickness of its outer envelope.

3.7 Integrate Eq. (3.45) and show that the time for a carbon white dwarf of mass $M$ to cool from a high internal temperature to a much lower internal temperature, $T_I$, is approximately

$$t = \frac{3}{5}\frac{kT_I}{L}\frac{M}{12m_{\mathrm{H}}},$$

where $L$ is the luminosity corresponding to $T_I$.

# CHAPTER 4

# Thermonuclear fusion in stars

Thermonuclear fusion in stars is activated by gravitational contraction. Because the fusion of nuclei is strongly hindered by Coulomb repulsion, the first nuclear fuel to ignite is composed of light nuclei with low charge. The energy released by this fuel brings a temporary halt to the contraction of the star. But contraction resumes when this particular fuel is exhausted. The internal temperature then rises until the next available fuel, consisting of heavier nuclei, is ignited. In this way a star can proceed through a sequence of nuclear burning stages which interrupt and delay gravitational contraction. These thermonuclear hang-ups not only prolong the life of a star, they also play a constructive role in the synthesis of heavier atomic nuclei. We shall begin this chapter by considering the basic physics of thermonuclear fusion.

## 4.1 THE PHYSICS OF NUCLEAR FUSION

The most remarkable aspect of thermonuclear fusion is that it happens at surprisingly low temperatures. Indeed, when the significance of nuclear fusion to stellar evolution was first noticed, many people expressed the doubt that stars were not hot enough for it to occur. But Sir Arthur Eddington's response to these doubters was robust:

> *We do not argue with the critic who urges that stars are not hot enough for this process; we tell him to go and find a hotter place.*

We now know that nuclear fusion in stars depends crucially on the wave-like properties of atomic nuclei.

**Barrier penetration**

Consider two nuclei with charges $Z_A$ and $Z_B$ with masses $m_A$ and $m_B$. At large separations $r$ these particles interact via a repulsive Coulomb potential $Z_A Z_B e^2/4\pi\epsilon_0 r$. However at distances comparable with a fermi, $10^{-15}$ m, they will also interact via a strong, attractive nuclear potential to give the overall potential energy of interaction shown schematically in Fig. 4.1. Note there is a Coulomb barrier which will inhibit the close approach of the nuclei and their fusion. The classical mechanics of a head-on collision is straightforward: the kinetic energy is progressively converted into potential energy as the nuclei approach each other until the kinetic energy falls to zero. They will then come momentarily to rest and bounce back. The distance of closest approach, $r_C$, corresponds to the point where the potential energy reaches the energy of approach. When this energy is $E$, $r_C$ is given by

$$E = \frac{Z_A Z_B e^2}{4\pi\epsilon_0 r_C}. \tag{4.1}$$

According to classical physics, fusion would only be possible if $r_C$ is less than $r_N$, the range of the nuclear interaction between the nuclei. In other words, fusion would only be possible if the nuclei have sufficient kinetic energy to climb over a Coulomb barrier of height

$$E_C = \frac{Z_A Z_B e^2}{4\pi\epsilon_0 r_N} \approx \frac{1.4\ Z_A Z_B}{(r_N \text{ in fermis})} \text{ MeV}. \tag{4.2}$$

The height of this barrier is large compared with the typical thermal energies of nuclei in stars. For example, when the temperature is $10^7$ K, $kT$ is of the order of a keV, not a MeV. Moreover, the fraction of nuclei with a thermal energy around a MeV is tiny; this fraction is of the order of $\exp(-E/kT)$, or $\exp(-1000)$ if $E = 1$ MeV and $kT = 1$ keV. Thus at first sight, Coulomb repulsion presents an insurmountable barrier to fusion in stars.

In fact, a definite distance of closest approach is a figment of the classical imagination, and fusion can occur at energies well below $E_C$. According to quantum mechanics, there is a chance that the nuclei can penetrate through the Coulomb barrier and reach the region where the strong nuclear interaction is effective. Once in this region, there is a possibility that the nuclei can fuse to form a heavier nucleus. The physics underlying barrier penetration is that the wave function representing the approach of the nuclei can leak into the region forbidden to classical particles. This wave function, $\psi(\mathbf{r})$, can be found by solving the Schrödinger equation for the two nuclei in the potential $V(r)$,

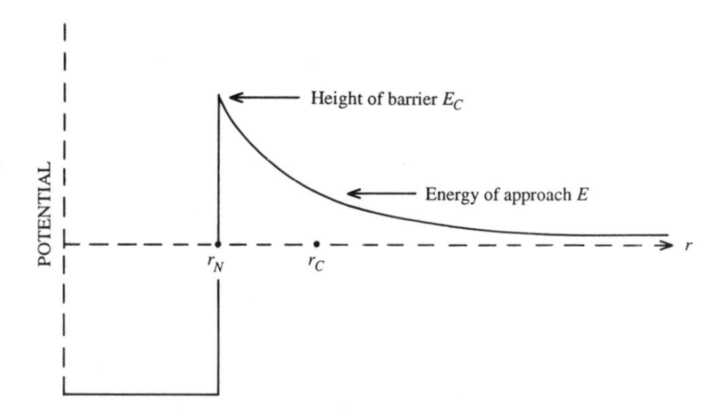


Fig. 4.1 A representation of the Coulomb and nuclear potentials between two nuclei of charge $Z_A$ and $Z_B$. The distance $r_C$ is the classical distance of closest approach for nuclei with an energy of approach equal to $E$. The distance $r_N$ represents the range of short-range nuclear forces. $E_C$ is the height of the Coulomb barrier keeping the nuclei apart.

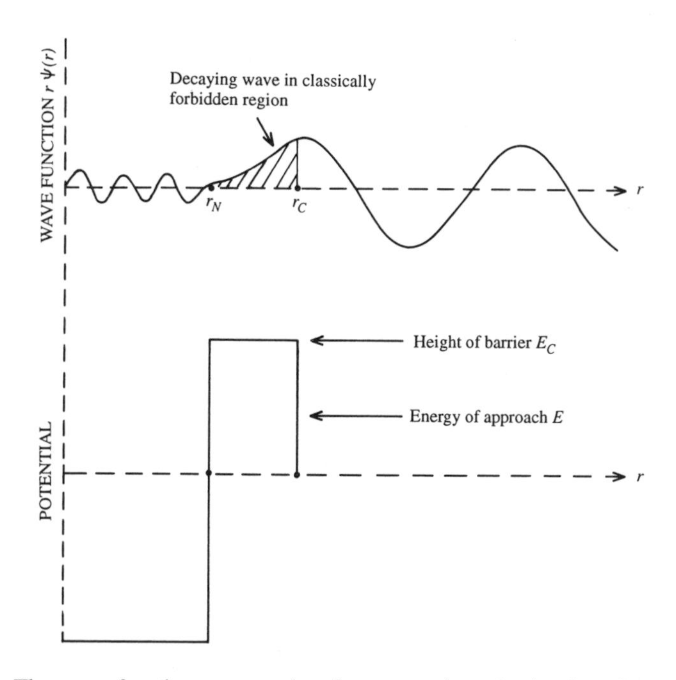


Fig. 4.2 The wave function representing the penetration of a barrier of constant height $E_C$ by particles whose energy of approach $E$ is below the barrier. The wave function, $r\psi(r)$, oscillates sinusoidally in the outer and inner classically allowed regions. It decays exponentially in the intervening classically forbidden region. In stellar thermonuclear fusion the wavelength for the relative motion of the nuclei in the outer classically allowed region is very long compared with the range of nuclear forces $r_N$.

$$\left[-\frac{\hbar^2\nabla^2}{2m_r} + V(r)\right]\psi(\mathbf{r}) = E\psi(\mathbf{r}), \tag{4.3}$$

where the reduced mass $m_r$ is given by $m_r = m_A m_B/(m_A + m_B)$. Once we know the wave function, we can find where the nuclei are likely to be found by noting that the probability that they are separated by a distance between $r$ and $r + \mathrm{d}r$ is $|\psi(\mathbf{r})|^2 4\pi r^2 \mathrm{d}r$.

To understand the wave mechanics of barrier penetration we consider the simple example of a barrier of constant height $E_C$, as shown in Fig. 4.2. When a particle approaches with energy $E$ from the right, the incoming wave function oscillates sinusoidally in the classically allowed region. As it penetrates into the classically forbidden region, the kinetic energy $E - E_C$ is negative and the incoming wave function satisfies the equation

$$\nabla^2\psi(\mathbf{r}) = \chi^2\psi(\mathbf{r}), \tag{4.4}$$

where $\chi$ is defined by

$$E = -\frac{\hbar^2\chi^2}{2m_r} + E_C. \tag{4.5}$$

It follows that the incoming wave function decays exponentially as $r$ gets smaller. In fact,

$$\psi(r) = \frac{\exp(\chi r)}{r}, \tag{4.6}$$

if there is no orbital angular momentum. The probability that the nuclei penetrate the Coulomb barrier is roughly given by

$$\text{Probability of penetration} \approx \frac{|\psi(r_N)|^2 4\pi r_N^2}{|\psi(r_C)|^2 4\pi r_C^2} = |\exp[-\chi(r_C - r_N)]|^2. \tag{4.7}$$

A more careful calculation would consider the probability current density of particles and reflection from the inner boundary, but the result given by Eq. (4.7) is adequate for our purposes.

This result may be adapted to give the probability of penetrating a barrier of variable height, such as the Coulomb barrier in Fig. 4.1. In this case the parameter $\chi$, which governs the exponential decay of the wave function in the classically forbidden region, depends on $r$. For nuclei with reduced mass $m_r$ we have

$$E = -\frac{\hbar^2[\chi(r)]^2}{2m_r} + \frac{Z_A Z_B e^2}{4\pi\epsilon_0 r}, \tag{4.8}$$

and the equation corresponding to Eq. (4.7) is

$$\text{Probability of penetration} \approx \left| \exp[- \int_{r_N}^{r_C} \chi(r)\mathrm{d}r] \right|^2. \tag{4.9}$$

The integral can be evaluated by substituting $r = r_C \cos^2\theta$. It is useful to write the result in terms of the relative energy $E$ of the nuclei and an energy $E_G$, called the Gamow energy, defined by

$$E_G = (\pi\alpha Z_A Z_B)^2 2m_r c^2, \tag{4.10}$$

where $\alpha$ is the dimensionless fine structure constant,

$$\alpha = \frac{e^2}{4\pi\epsilon_0 \hbar c} \approx \frac{1}{137}. \tag{4.11}$$

Equation (4.9) then leads to

$$\text{Probability of penetration} \approx \exp\left[ - \left( \frac{E_G}{E} \right)^{1/2} \right]. \tag{4.12}$$

Thus, the Coulomb barrier keeping charged nuclei apart need not be overcome in order to give the nuclei a chance to fuse. In practice, stars evolve slowly by adjusting their temperature so that the average thermal energy of nuclei is well below the Coulomb barrier. Fusion then proceeds at a rate proportional to the probability of penetration of the barrier. Because this probability is very low, fusion proceeds at a slow pace and the nuclear fuel lasts for an astronomically long time. We note that the penetrability of the barrier is completely described by its Gamow energy, Eq. (4.10). For the fusion of two protons $E_G$ is 493 keV. If the temperature is about $10^7$ K, the typical thermal energy, $kT$, is about 1 keV, and the penetration probability for two protons with this typical energy is $\exp[-(E_G/kT)^{1/2}] \approx \exp[-22]$. There are, of course, protons present with higher kinetic energy which will have a better chance of penetrating the Coulomb barrier.

## Fusion cross-sections

The probability of fusion is usually expressed in terms of a fusion cross-section. In order to define a cross-section for a particular reaction, we consider a particle passing through a medium containing $n$ target particles per unit volume. The probability that the incoming particle reacts as it travels an infinitesimal distance $\Delta x$ is defined by

$$\text{Probability of reaction in distance } \Delta x = \sigma n \Delta x, \tag{4.13}$$

where $\sigma$ is the reaction cross-section. It follows that the probability of no reaction in $\Delta x$ is $[1 - \sigma n \Delta x]$. The probability that the particle travels a finite distance $x$ without reaction can be found by dividing the distance $x$ into $N$ intervals of thickness $\Delta x = x/N$, and then compounding the probabilities for no reaction in each of the intervals. We have:

$$\text{Probability of no reaction in distance } x = \lim_{N\to\infty} [1 - \sigma n x/N]^N = \exp[-\sigma n x]. \quad (4.14)$$

The mean free path travelled before a reaction is then given by

$$\bar{l} = \int_0^\infty x \exp[-\sigma n x]\, \sigma n \, \mathrm{d}x = \frac{1}{n\sigma}. \quad (4.15)$$

The reaction cross-section, $\sigma$, is an effective target area which is proportional to the probability of the reaction occurring in a collision. A classical analogue is the collision between a cricket ball and a window of area 1 $m^2$; the reaction cross-section would be 0.1 $m^2$ if there is a 10% chance that the window breaks. A nuclear cross-section depends on the energy of the nuclei and their electromagnetic and nuclear interactions. In particular, the cross-section often exhibits resonant behaviour when the energy matches the energy needed to form a compound nuclear state. The unit usually used for nuclear cross-sections is the barn which equals $10^{-28}$ $m^2$; a particularly large nuclear cross-section, will be *as big as a barn*, but millibarn and microbarn cross-sections are more common.

Our chief concern is the cross-section for the fusion of two nuclei. At low energies this cross section is proportional to the probability of penetration of the Coulomb barrier keeping the nuclei apart. We therefore use Eq. (4.12) and write the fusion cross-section for nuclei with relative energy $E$ as

$$\sigma(E) = \frac{S(E)}{E} \exp\left[-\left(\frac{E_G}{E}\right)^{1/2}\right]. \quad (4.16)$$

The energy dependence of the fusion cross-section is invariably dominated by a steeply rising probability of barrier penetration. The factor $S(E)$, which is determined by the nuclear physics of fusion, varies much more slowly with energy; sometimes, however, it may peak when the energy is near a nuclear resonance. The factor of $1/E$ has been introduced because nuclear cross-sections at low energies are often proportional to the square of the de Broglie wavelength for the relative motion of the nuclei before fusion; if $p$ is the relative momentum of the nuclei, $\lambda^2 = h^2/p^2 = h^2/2m_r E$.

In practice, it is very difficult to measure fusion cross-sections at energies directly relevant to astrophysics; i.e. at energies well below the Coulomb barrier. Larger, and hence more easily observed cross-sections are measured at higher energies, and the prescription for the fusion cross-section given by Eq. (4.16) is used to extrapolate the data to lower, more relevant energies. This prescription is also very useful in calculating the temperature dependence of thermonuclear reaction rates.

**Thermonuclear reaction rates**

Consider a hot ionized gas containing nuclei of type $A$ and $B$ with concentrations $n_A$ and $n_B$ which can fuse with a fusion cross-section denoted by $\sigma$. For the moment we will neglect the motion of the $B$ nuclei and assume that all the $A$ nuclei move with speed $v$. According to Eq. (4.15), a nucleus of type $A$ travels an average a distance of $1/n_B\sigma$ before fusing with a $B$ nucleus, and the average time before fusion is $\tau_A = 1/n_B\sigma v$. Thus in unit volume of the gas, we have $n_A$ nuclei of type $A$ which fuse at a rate of $R_{AB} = n_A n_B \sigma v$ per second.

Of course, both types of nuclei move and the fusion cross-section depends on the relative speed $v_r$ of the nuclei. If $P(v_r)\,\mathrm{d}v_r$ denotes the probability that the relative speed is between $v_r$ and $v_r+\mathrm{d}v_r$, then the average value of the product of the fusion cross-section and the relative speed is

$$\langle \sigma v_r \rangle = \int_0^\infty \sigma v_r P(v_r)\,\mathrm{d}v_r. \tag{4.17}$$

When we take this averaging procedure into account, the mean time for a particular nucleus of type $A$ to fuse with a $B$ nucleus becomes

$$\tau_A = \frac{1}{n_B\langle \sigma v_r \rangle}, \tag{4.18}$$

and the $A - B$ fusion rate per unit volume becomes

$$R_{AB} = n_A n_B \langle \sigma v_r \rangle. \tag{4.19}$$

Care is needed in using these equations to describe the fusion of identical nuclei. The product $n_A n_B$ in Eq. (4.19) represents the number of possible pairs of nuclei that can fuse. It must be replaced by $n(n-1)/2$, or in practice $n^2/2$, when calculating the fusion rate for identical nuclei with concentration $n$.

In most astrophysical situations the nuclei form a classical, non-relativistic gas with a speed distribution given by the Maxwell–Boltzmann distribution. Further, it is easy to show that Maxwellian distributions for nuclei $A$ and $B$ lead to a Maxwellian distribution for the relative speed, given by

$$P(v_r)\mathrm{d}v_r = \left[\frac{m_r}{2\pi kT}\right]^{3/2} \exp\left[-\frac{m_r v_r^2}{2kT}\right] 4\pi v_r^2 \mathrm{d}v_r. \tag{4.20}$$

If this distribution is substituted into Eq. (4.17) and if the integration variable is changed to the energy $E = \frac{1}{2}m_r v_r^2$, we obtain

$$\langle \sigma v_r \rangle = \left[\frac{8}{\pi m_r}\right]^{1/2} \left[\frac{1}{kT}\right]^{3/2} \int_0^\infty E\sigma(E) \exp\left[-\frac{E}{kT}\right] \mathrm{d}E. \tag{4.21}$$

If we substitute the prescription Eq. (4.16) for the fusion cross-section into Eq. (4.21) and use Eq. (4.19), we obtain the following expression for the thermonuclear fusion rate per unit volume

$$R_{AB} = n_A n_B \left[\frac{8}{\pi m_r}\right]^{1/2} \left[\frac{1}{kT}\right]^{3/2} \int_0^\infty S(E) \exp\left[-\frac{E}{kT} - \left(\frac{E_G}{E}\right)^{1/2}\right] \mathrm{d}E. \tag{4.22}$$

Note that in order to react at energy $E$, the nuclei need to borrow an energy $E$ from the thermal environment, and the probability of a successful loan is proportional to the Boltzmann factor $\exp[-E/kT]$ in Eq. (4.22). Moreover in order to fuse, the nuclei must first penetrate the Coulomb barrier keeping them apart, and the probability of penetration is given by the factor $\exp[-(E_G/E)^{1/2}]$ in Eq. (4.22). Once this has happened, nuclear forces can sometimes bring about a fusion. The nuclear physics of fusion is hidden in the factor $S(E)$ in Eq. (4.22).

Because the nuclear factor $S(E)$ usually varies slowly with energy, the energy dependence of the integrand in Eq. (4.22) is governed by the exponential borrowing and penetrating functions. As illustrated in Fig. 4.3, the product of these two exponentials has a maximum when the energy $E$ is equal to

$$E_0 = \left[\frac{E_G(kT)^2}{4}\right]^{1/3}. \tag{4.23}$$

Figure 4.3 also indicates that fusion dominantly takes place in a narrow energy range around a most likely fusion energy equal to $E_0$. We can find the width of this window for fusion by making a Taylor's expansion about $E_0$ to give the following approximation:

$$\exp\left[-\frac{E}{kT} - \left(\frac{E_G}{E}\right)^{1/2}\right] \approx \exp\left[-3\left(\frac{E_G}{4kT}\right)^{1/3}\right] \exp\left[-\left(\frac{(E-E_0)}{\Delta/2}\right)^2\right], \tag{4.24}$$

where $\Delta$, the width of the fusion window, is given by

$$\Delta = \frac{4}{3^{1/2}2^{1/3}} E_G^{1/6}(kT)^{5/6}. \tag{4.25}$$

These expressions for $E_0$ and $\Delta$ show that fusion mostly occurs at energies determined by the temperature of the gas and the Gamow energy of the Coulomb

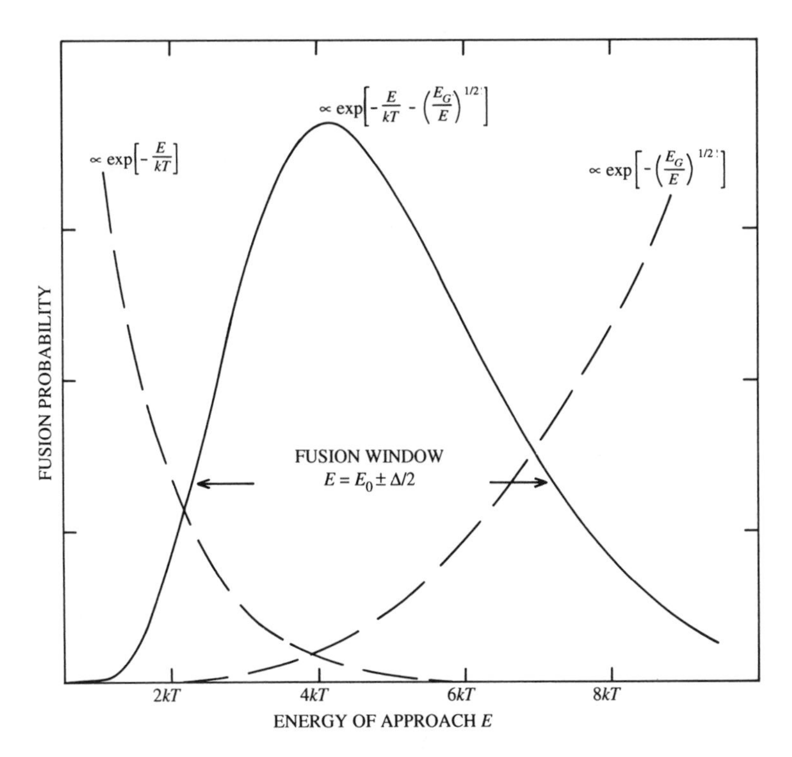


Fig. 4.3 The energy window for the fusion of nuclei with a Gamow energy $E_G$ and temperature $T$. To react at energy $E$, the nuclei need to borrow an energy $E$ from the thermal environment, and the probability of a successful loan is proportional to the Boltzmann factor $\exp[-E/kT]$. To fuse, the nuclei must first penetrate the Coulomb barrier keeping them apart, and the probability of penetration is given by the factor $\exp[-(E_G/E)^{1/2}]$. The product of these two factors indicates that fusion mostly occurs in an energy window $E_0 \pm\Delta/2$. For the fusion of two protons at $2 \times 10^7$ K, $E_G = 290kT$, $E_0 = 4.2kT$, and $\Delta = 4.8kT$, as illustrated.

barrier. We recall that the Gamow energy is simply related to the charges of the nuclei and their reduced mass via Eq. (4.10), namely

$$E_G = (\pi\alpha Z_A Z_B)^2 2m_r c^2.$$

For example, the Gamow energy for two protons is $E_G = 493$ keV. When the temperature is $2 \times 10^7$ K, $kT = 1.7$ keV and the fusion of two protons is most likely at $E_0 = 7.2$ keV. The half-width of the fusion window, $\Delta/2$, is 4.1 keV.

In many cases the nuclear factor $S(E)$ is approximately constant across the fusion window. It can then be replaced by a constant $S(E_0)$ , and Eq. (4.22) simplifies to

$$R_{AB} = n_A n_B \left[\frac{8}{\pi m_r}\right]^{1/2} \left[\frac{1}{kT}\right]^{3/2} S(E_0) \int_0^{\infty} \exp\left[-\frac{E}{kT} - \left(\frac{E_G}{E}\right)^{1/2}\right] \mathrm{d}E. \quad (4.26)$$

The value of the integral in this equation may be evaluated using the approximation (4.24) to give $\sqrt{\pi}\Delta/2$ times the maximum value of the integrand at $E = E_0$. When the numerical values for the various constants are inserted we find a fusion rate given by

$$R_{AB} = 6.48 \times 10^{-24} \frac{n_A n_B}{A_r Z_A Z_B} S(E_0) \left(\frac{E_G}{4kT}\right)^{2/3} \exp\left[-3\left(\frac{E_G}{4kT}\right)^{1/3}\right] \quad \mathrm{m}^{-3}\mathrm{s}^{-1}, \quad (4.27)$$

where $A_r$ is the reduced mass of the nuclei in atomic mass units and $S(E_0)$ is the nuclear fusion factor in units of keV barns.

We now have a three parameter model for thermonuclear fusion. The parameters are the nuclear fusion factor $S(E_0)$, the Gamow energy $E_G$, and the temperature $T$. The nuclear factor depends on the specific nuclear reaction taking place; in practice, it is usually measured in accelerator experiments. The Gamow energy depends simply on the charge of the nuclei and their reduced mass and, together with the temperature, it determines how the Coulomb barrier affects thermonuclear fusion.

The main effects of the Coulomb barrier can be identified by focusing on the terms involving exponential dependence on $E_G$ and $T$. If we do this, we find that the key factors in the expression for the fusion rate are

$$R_{AB} \propto n_A n_B S(E_0) \exp\left[-3\left(\frac{E_G}{4kT}\right)^{1/3}\right]. \quad (4.28)$$

The exponential term in Eq. (4.28) can be thought of as a slow-down factor due to the Coulomb barrier. It clearly demonstrates why, if there are many species present, there is a strong tendency for those with the smaller Coulomb barrier to take part in thermonuclear fusion more rapidly. As an example, we compare

$$p + d \rightarrow {}^3\mathrm{He} + \gamma$$

and

$$p + {}^{12}\mathrm{C} \rightarrow {}^{13}\mathrm{N} + \gamma.$$

The Gamow energies for these two fusion reactions are 0.657 MeV and 35.5 MeV, respectively. At a temperature of $2 \times 10^7$ K these fusion reactions are slowed down by factors of exp(−14) and exp(−52), respectively.

The exponential term in Eq. (4.28) also demonstrates that the fusion rate increases rapidly with temperature. Indeed, Eq. (4.28) implies that

$$\frac{\mathrm{d}R_{AB}}{\mathrm{d}T} \approx \left[\frac{E_G}{4kT}\right]^{1/3} \frac{R_{AB}}{T}. \tag{4.29}$$

For the fusion of protons and deuterons at a temperature near $2 \times 10^7$ K

$$\frac{\mathrm{d}R_{pd}}{\mathrm{d}T} \approx 4.6 \frac{R_{pd}}{T}. \tag{4.30}$$

This implies that as $T$ varies about $2 \times 10^7$ K, the proton–deuteron fusion rate varies as $T^{4.6}$. In fact, when account is taken of the factor of $T^{-2/3}$ in Eq. (4.27), the fusion rate is approximately proportional to $T^4$. The temperature dependence is more marked for fusion reactions with a higher Coulomb barrier. For the fusion of protons with $^{12}$C nuclei, the fusion rate near $2 \times 10^7$ K is proportional to $T^{17}$.

Even though the Coulomb barrier plays a dominant role in shaping the properties of all thermonuclear reactions, the actual rates depend on the *interactions* that bring about the fusion. Nuclear strong, electromagnetic and nuclear weak interactions may be involved. The net effect is summarized by the nuclear factor $S(E)$. We shall consider particular thermonuclear reactions later in this chapter. But at this stage it is useful to note that $S(E)$ is necessarily small if the reaction relies on the nuclear weak interaction; such reactions involve the emission of a neutrino. It is larger for reactions which are reliant on the electromagnetic interaction and emit photons. It is larger still for reactions governed by the nuclear strong interaction.

## 4.2 HYDROGEN BURNING

Star formation begins with the gravitational collapse of a cloud composed of hydrogen, helium and traces of other chemical elements. The collapse is rapid until the atoms are ionized and energy can no longer easily escape from the cloud. The cloud then contracts slowly in a state close to hydrostatic equilibrium; half the gravitational energy released is lost as radiation and the other half heats up the cloud. This contraction will continue until the activation of a source of energy other than gravity. The first such source, which is activated when the temperature is about $10^6$ K, is the thermonuclear fusion of protons with light nuclei, such as D, Li, Be and B. This involves fast, indeed bomb-like, reactions. But only a limited amount of energy is released because the light nuclei are only present in small quantities and are rapidly consumed. In order to properly begin its life as a star, the hot ionized gas must find some way of exploiting the nuclear fuel provided by its dominant nuclear constituent, protons. It must find a way of burning ordinary hydrogen.

The net effect of hydrogen burning is to transform protons to $^4$He nuclei. We note that protons must be converted into neutrons at some stage during a chain of reactions which burn hydrogen, and that this transformation can only be effected by a nuclear weak process. The most likely process is $p \to n + e^+ + \nu_e$ and, in this case, the net result of a hydrogen-burning chain is

$$4p \to {}^4\text{He} + 2e^+ + 2\nu_e. \tag{4.31}$$

The decrease in mass in this transformation implies a kinetic energy release of 24.69 MeV. But each of the positrons will promptly annihilate with an electron and release a further $2m_ec^2 = 1.02$ MeV to give total energy release of 26.73 MeV. However, a small percentage of this energy is associated with the kinetic energy of the neutrinos. This is not retained locally but escapes almost without interaction.

Hydrogen burning would be a straightforward and rapid process, if a bound state of two protons existed. Such a state would be an isotope of helium, $^2$He, and hydrogen would begin to burn via the electromagnetic reaction

$$p + p \to {}^2\text{He} + \gamma,$$

and each $^2$He would then beta decay to form a deuteron. But the nuclear force between two protons is not quite strong enough to produce a $^2$He bound state. Indeed, the absence of a $^2$He bound state implies that hydrogen burning is a subtle and slow process. In fact, as first explained by Bethe in 1939, there are two main ways of burning hydrogen, the proton–proton chain and the carbon–nitrogen cycle.

### The proton–proton chain

One sure, but slow, way of by-passing the bottle-neck formed by the absence of a $^2$He bound state is to fuse protons via a weak nuclear reaction:

$$p + p \to d + e^+ + \nu_e. \tag{4.32}$$

The underlying mechanism for this reaction is that one of the interacting protons undergoes inverse beta decay, $p \to n + e^+ + \nu_e$, and the neutron produced is then bound to the other proton to form a deuteron. The first step in this mechanism is a virtual process because an energy of least 1.8 MeV is needed to convert a proton into a neutron, a positron and a massless neutrino. But this energy is more than paid back by the formation of a deuteron with a binding energy of 2.225 MeV.

However, the key feature of the reaction (4.32) is that it is very slow. Indeed, this reaction is so slow that it has not been possible to measure its cross-section in an experiment on earth. But according to theoretical calculations, it has a nuclear $S$ factor of $S_{pp}(0) = 3.8 \times 10^{-22}$ keV barns. The mean lifetime of a proton before fusion and the proton–proton fusion rate in stellar material can then be found using this value for $S_{pp}$ and Eqs. (4.18), (4.19) and (4.27). We can estimate the fusion rate at the centre of the sun by taking $T = 15 \times 10^6$ K, $\rho = 10^5$ kg m$^{-3}$ and a hydrogen mass fraction $X_1 = 0.5$. The concentration of protons is then $n_p = X_1\rho/m_{\rm H} = 3 \times 10^{31}$ m$^{-3}$ and the proton–proton fusion rate is $5 \times 10^{13}$ s$^{-1}$ m$^{-3}$. This implies that a proton in the centre of the sun has to hang around for about $9 \times 10^9$ years on average before it fuses with another proton. This astronomically long time sets the time scale for the hydrogen burning phase of the sun's life.[1]

Once deuterons are formed by the reaction (4.32), the way is open for much faster reactions to synthesize $^4$He nuclei. There are three sequences of reactions which form the main branches of the proton–proton chain. These branches, labelled by I, II, and III, are shown in Fig. 4.4.

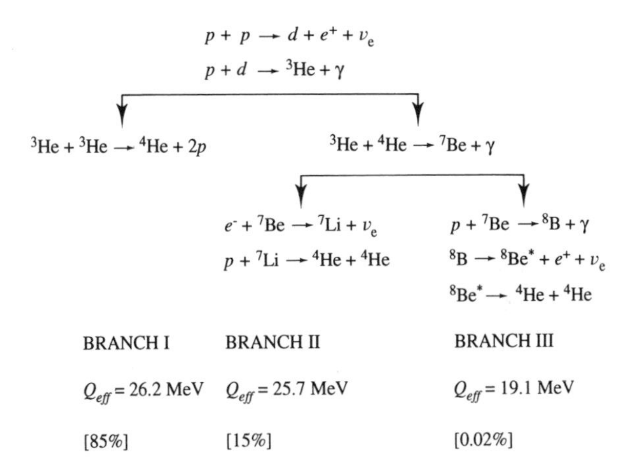


Fig. 4.4 The three competing branches of the proton–proton chain with the net result $4p \to {}^4\text{He} + Q_{eff}$. Here $Q_{eff}$ is the effective energy released by the branch; it includes the energy from the annihilation of positrons, but it does not include any of the energy carried away by neutrinos. Note, a pre-existing $^4$He nucleus acts as a catalyst in branches II and III, its destruction leading to two new $^4$He nuclei. According to the Standard Solar Model, Bahcall (1989), the proton–proton chain in the sun is terminated by branch I 85% of the time, by branch II 15% of the time and by branch III 0.02% of the time.

[1] It is sometimes suggested that the time scale for hydrogen burning would be shorter if it was initiated by an electromagnetic reaction instead of the weak-nuclear reaction (4.32). This is not the case, because the overall rate for hydrogen burning is determined by the rate at which energy can escape from the star, i.e. by its opacity. If hydrogen burning was initiated by an electromagnetic reaction, this reaction would proceed at about the same rate as the weak reaction (4.32), but at lower temperature and density.

A deuteron formed by reaction (4.32) is almost immediately snapped up by the second reaction in the proton–proton chain

$$p + d \rightarrow {}^3\text{He} + \gamma. \tag{4.33}$$

The $S$ factor for this electromagnetic reaction is $S_{pd}(0) = 2.5 \times 10^{-4}$ keV barns, 18 orders of magnitude greater than the $S$ factor for the nuclear weak reaction (4.32). As a result, a deuteron in the centre of the sun lives for about a second before it fuses with a proton. Note that, since deuterons are produced at a slow rate and consumed at a fast rate, the equilibrium abundance of deuterons is expected to be low. If we apply Eq. (4.19) we find that the rate of change in the concentration of deuterons is

$$\frac{dn_d}{dt} = \frac{1}{2} n_p^2 \langle \sigma v_r \rangle_{pp} - n_p n_d \langle \sigma v_r \rangle_{pd}. \tag{4.34}$$

Hence the deuteron concentration will increase and reach an equilibrium concentration given by

$$\frac{n_d}{n_p} = \frac{\langle \sigma v_r \rangle_{pp}}{2 \langle \sigma v_r \rangle_{pd}} \approx \frac{S_{pp}(0)}{2 S_{pd}(0)}. \tag{4.35}$$

Thus the deuteron to proton ratio in the centre of a star like the sun is determined by the ratio of a nuclear weak cross-section to an electromagnetic cross-section. This ratio is tiny, of the order of $10^{-18}$. In contrast, terrestrial deuterium is relatively abundant; about 0.015 % of the hydrogen atoms around you are deuterium atoms. It is clear that terrestrial deuterium cannot be a product of thermonuclear reactions in stars like the sun. It was, in fact, produced during the very early universe, minutes after the big bang.

A $^3$He nucleus formed by reaction (4.33), can be processed in the two ways shown in Fig. 4.4. It can either complete the branch I of the chain by fusing with another $^3$He nucleus, or it can fuse with a $^4$He nucleus. The latter alternative leads to the formation of $^7$Be, which can be processed in two ways, and to the termination of the proton–proton chain via branches II and III, as shown in Fig. 4.4. Note that a pre-existing $^4$He nucleus acts as a catalyst when the proton–proton chain is terminated by branch II or III; it is destroyed when it fuses with a $^3$He nucleus but two more $^4$He nuclei are formed subsequently.

The proton–proton fusion reaction, Eq. (4.32), is the first and slowest link in the proton–proton chain, and as such it governs the rate at which energy is released by the chain as a whole. This rate is simply the proton–proton fusion rate, $R_{pp}$, times the energy released by the chain per proton–proton fusion. Note, however, that two proton–proton fusions are needed to produce a $^4$He nucleus via branch I, but only one is needed if the chain is completed via branches II or III. Hence the energy release by the chain per proton–proton fusion is sensitive to the relative importance of the three branches of the chain. According to the Standard Solar Model, Bahcall (1989), the proton–proton chain in the sun is terminated by branch I 85% of the time, by branch II 15% of the time and by branch III 0.02% of the time. It follows that the average energy released per proton–proton fusion in the sun is

$$0.85 \times 26.2/2 + 0.15 \times 25.7 = 15 \text{ MeV}. \tag{4.36}$$

If we combine this with our earlier estimate of $5 \times 10^{13}$ m$^{-3}$ s$^{-1}$ for the proton–proton fusion rate in the solar centre, we find an energy production rate of about 120 W m$^{-3}$.

Finally, it is useful to have an approximate expression for the energy production rate which clearly indicates how it depends on the temperature, density and mass fraction of hydrogen. The temperature dependence of the proton–proton fusion rate can be found by using the appropriate Gamow energy, $E_G$ = 493 keV, and Eq. (4.27); for temperatures close to $T = 15 \times 10^6$ K, the typical temperature at the centre of the sun, the fusion rate $R_{pp}$ is approximately proportional to $T^4$. This rate is also proportional to $n_p^2$ or to $X_1^2\rho^2$. Hence, the energy production rate by the proton–proton chain is proportional to $X_1^2\rho^2T^4$. If we normalize to an energy production rate of 120 W m$^{-3}$ at $T = 15 \times 10^6$ K, $\rho = 10^5$ kg m$^{-3}$ and $X_1 = 0.5$, we find that the proton–proton chain produces energy at a rate given by

$$\varepsilon_{pp} = 9.5 \times 10^{-37}\, X_1^2\rho^2T^4 \text{ W m}^{-3}. \tag{4.37}$$

### The carbon–nitrogen cycle

The proton–proton chain can account for hydrogen burning in main sequence stars with masses comparable to the sun, but it fails in more massive stars. Even though the internal temperatures of such stars are only moderately higher than the sun's, their luminosities are much higher, too high to be explained by the $T^4$ dependence of the proton–proton chain. We recall that this temperature dependence is governed by the Coulomb barrier between two protons; see Eq. (4.29). To explain the luminosities of massive main sequence stars, such as Sirius A, a more temperature-dependent mechanism for hydrogen burning is needed, a mechanism which must be governed by a higher Coulomb barrier. Such a mechanism must involve heavy elements. But, because these elements are, at best, present in low abundance, recycling of this scarce resource is needed to prolong the hydrogen burning. In fact, stars can burn hydrogen by an almost closed sequence of reactions called the carbon–nitrogen cycle. This cycle depends on the presence of small quantities of carbon, carbon produced in earlier generations of stars by helium burning; this carbon is not completely destroyed but partially converted into nitrogen. The reactions of the carbon–nitrogen cycle are illustrated in Fig. 4.5.

REACTIONS OF THE CARBON–NITROGEN CYCLE

$p + {}^{12}C \rightarrow {}^{13}N + \gamma$ [S (0) = 1.5 keV barns]
$\quad \hookrightarrow {}^{13}C + e^+ + \nu_e$
$p + {}^{13}C \rightarrow {}^{14}N + \gamma$ [S (0) = 5.5 keV barns]

$p + {}^{14}N \rightarrow {}^{15}O + \gamma$ [S (0) = 3.3 keV barns]
$\quad \hookrightarrow {}^{15}N + e^+ + \nu_e$
$p + {}^{15}N \rightarrow {}^{12}C + {}^{4}He$ [S (0) = 78 keV barns]

Fig. 4.5 Hydrogen burning by the carbon–nitrogen cycle. The net result of this sequence of reactions is $4p \rightarrow {}^4He + Q_{eff}$. The effective energy released $Q_{eff}$ is 23.8 MeV; this includes the energy from the annihilation of positrons, but it does not include the energy carried away by neutrinos. Note that nuclei of carbon and nitrogen are temporarily transformed but return to take part in subsequent operations of the cycle. The rates for these reactions are governed by the relevant Coulomb barriers and the approximate $S$ factors indicated.

Even though the carbon–nitrogen cycle has no beginning or end, it is useful to think of it as commencing with the capture of a proton by a ${}^{12}C$ nucleus. This is followed by a transformation of a proton into a neutron by a beta decay, the capture of two more protons, the transformation of a second proton into a neutron by another beta decay and finally the capture of a fourth proton to produce a new ${}^{12}C$ nucleus and a ${}^4He$ nucleus. The net effect of this sequence is the transformation $4p \rightarrow {}^4He + 2\ e^+ + 2\nu_e$, and a ${}^{12}C$ nucleus still in circulation. Thus, carbon acts as a catalyst for hydrogen burning. In fact other cycles also exist, particularly one involving ${}^{16}O$. However the carbon-nitrogen cycle illustrated in Fig. 4.5 is by far the most important.

The rate of energy production by the carbon–nitrogen cycle is governed by the slowest reaction in the sequence. By considering the Coulomb barriers and the values of the $S$ factors given in Fig. 4.5, we conclude that the slowest reaction is

$$p + {}^{14}N \rightarrow {}^{15}O + \gamma. \tag{4.38}$$

The mean life for a ${}^{14}N$ nucleus in the centre of the sun can be estimated by assuming a density of $10^5$ kg m$^{-3}$, a temperature of $15 \times 10^6$ K, a mass fraction of hydrogen of 0.5 and a nuclear $S$ factor of 3.3 keV barns. Substitution into Eqs. (4.18), (4.19) and (4.27) shows that a ${}^{14}N$ nucleus in the sun has an average life of about $5 \times 10^8$ years before it fuses with a proton. The fusion rate per unit volume depends on the concentration of nitrogen nuclei in the solar centre. In the Standard Solar Model, the abundance of ${}^{14}N$ at the solar centre is about 0.6% which implies a concentration of nitrogen nuclei of about $0.006\rho/14m_H = 2.6 \times 10^{28}$ m$^{-3}$. As each of these nuclei lasts for an average of $5 \times 10^8$ years, the fusion rate is approximately $1.6 \times 10^{12}$ m$^{-3}$ s$^{-1}$. In contrast, protons fuse via reaction (4.32) at a much faster rate of $5 \times 10^{13}$ m$^{-3}$ s$^{-1}$. We conclude that the carbon–nitrogen cycle is not an important source of energy production in the sun. Indeed, accurate calculations show that 98.4% of the solar energy is due to the proton–proton chain and only 1.6% is due to the carbon–nitrogen cycle.

However, the high Coulomb barriers involved in the carbon–nitrogen cycle imply fusion rates which increase very rapidly with temperature. In particular, if we calculate the Gamow energy corresponding to reaction (4.38) and use Eq. (4.27), we find a fusion rate proportional to $T^{18}$. The energy production via the carbon–nitrogen cycle has a similar temperature dependence, which is much more rapid than the $T^4$ dependence of the proton–proton chain. Thus, we expect the carbon–nitrogen cycle to be the dominant source of energy production in massive main sequence stars which burn hydrogen at temperatures higher than the central temperature of the sun.

Finally we note that the carbon–nitrogen cycle has an important role in stellar nucleosynthesis. It not only transforms hydrogen to helium, it also transforms $^{12}$C, made by helium burning in a star of an earlier generation, into $^{13}$C, $^{14}$N, and $^{15}$N. Indeed, if we assume equilibrium conditions and if we neglect leakage from the cycle, the relative abundances of these nuclei are inversely proportional to their fusion rates. For example, in the centre of a star burning hydrogen by the carbon-nitrogen cycle at a temperature of $50 \times 10^6$ K, the relative abundances of $^{12}$C, $^{13}$C, $^{14}$N, and $^{15}$N are 4%, 1%, 95% and 0.004%, respectively. The high abundance of $^{14}$N arises because its fusion rate is the slowest in the cycle. In fact, the vitally important nitrogen in the solar system is a product of hydrogen burning by the carbon–nitrogen cycle in earlier generations of nearby stars. Other elements were also produced by other cycles of reactions which couple with the carbon–nitrogen cycle.

### Solar neutrinos

There is no question that hydrogen burning, mostly via the proton–proton chain and the carbon–nitrogen cycle, provides a viable power source for main sequence stars. But few of the details are open to observation because the whole process is obscured by millions of kilometres of stellar matter. But the cover-up is not complete. Neutrinos created by hydrogen burning can escape almost without interaction straight from the heart of a star. In so doing, they carry inside information on what is actually happening at the centre. Detection of these neutrinos is clearly a formidable task. Not only do they hardly interact as they escape from the star, they hardly interact when they arrive at the earth. Nevertheless, neutrinos from the nearest star, the sun, were first detected in 1968 in a pioneering experiment set up by R. Davis in the Homestake Gold Mine in South Dakota. This experiment has been developed and data has been taken for over 20 years. The long-standing problem of reconciling the results of this experiment with the theoretical predictions is called *the solar neutrino problem*.

Hydrogen burning necessarily involves the emission of neutrinos. They arise when the nuclear weak interaction changes a proton to a neutron via $p \rightarrow n + e^{+} + \nu_e$. This must occur twice during the hydrogen burning process $4p \rightarrow {}^4\text{He} + 2e^{+} + 2\nu_e$. The expected flux of neutrinos can be found by noting that the formation of each $^4$He is accompanied by the release of two neutrinos and a thermal energy $Q_{eff}$, which according to Figs 4.4 and 4.5 is about 26 MeV. Hence if hydrogen burning is the power source for the sun's luminosity of $L_{\odot} = 3.86 \times 10^{26}$ W, neutrinos must be released at a rate of $2L_{\odot}/Q_{eff} = 1.86 \times 10^{38}\ \text{s}^{-1}$.

To escape from the sun, each neutrino must travel a distance of about $R_{\odot}$, i.e. $7 \times 10^8$ m. The probability of interaction during this escape is $\sigma n R_{\odot}$, where $\sigma$ is the average interaction cross-section with an electron or a nucleus and $n$ is the average density of electrons and nuclei in the sun. Since $\sigma$ is of the order of $10^{-48}\ \text{m}^2$ and $n$ is approximately $10^{30}\ \text{m}^{-3}$, the probability of interaction is an insignificant $10^{-9}$. Thus, neutrinos do indeed escape almost unhindered from the sun and arrive some eight minutes later at the earth, a distance of $1.5 \times 10^{11}$ m away. The neutrino flux at the earth is $F_{\nu} = 6.6 \times 10^{14}\ \text{m}^{-2}\ \text{s}^{-1}$, an intense but almost undetectable flow straight from the heart of the sun.

This neutrino flux is the combined effect of a number of reactions and decays in the proton–proton chain and in the carbon–nitrogen cycle. These processes may be identified by inspecting Figs 4.4 and 4.5. Clearly each process emits neutrinos with an energy spectrum characteristic of the process, but with a rate which depends on the details of hydrogen burning inside the sun. The most accurate predictions for these rates are based upon a detailed model of the sun, called the Standard Solar Model. This model, which is described by Bahcall in his book *Neutrino Astrophysics*, is really an evolutionary sequence of models. The sequence begins with a star with a homogeneous composition similar to that observed on the solar surface, and successive models are then calculated by allowing for the changes in composition brought about by hydrogen burning. This sequence is required to fit the known data, namely the age, the mass, the radius, the surface composition and the present-day luminosity.

The predictions of the Solar Standard Model for the flux of neutrinos from various processes taking place inside the sun are given in Table 4.1. As expected, the majority of the neutrinos originate from the primary reaction of the proton–proton chain, but these neutrinos have low energy, never exceeding 0.420 MeV. The neutrinos from electron capture by $^7$Be, the reaction which initiates branch II of the proton–proton chain, are the next most plentiful. The flux of neutrinos from $^8$B decay in branch III of the chain is four orders of magnitude less, but these neutrinos are very energetic. In addition, there are contributions from $^{13}$N and $^{15}$O beta decay, two processes in the carbon–nitrogen cycle. These contributions are small because the carbon–nitrogen cycle only supplies 1.6% of the solar luminosity. When combined, the fluxes from individual processes should yield a net flux of $F_{\nu} = 6.6 \times 10^{14}\ \text{m}^{-2}\ \text{s}^{-1}$ at the earth.

TABLE 4.1 The flux of neutrinos from particular processes in the proton–proton chain and the carbon–nitrogen cycle inside the sun. In addition there are small contributions from $p + e^- + p \rightarrow d + \nu_e$ and $p + {}^3\text{He} \rightarrow {}^4\text{He} + e^+ + \nu_e$; see Bahcall (1989) for details.

| Process | Neutrino flux ($10^{14}$ m$^{-2}$ s$^{-1}$) | Maximum neutrino energy (MeV) |
|---|---|---|
| $p + p \rightarrow d + e^+ + \nu_e$ | 6.0 (1 ± 0.02) | 0.420 |
| $e^- + {}^7\text{Be} \rightarrow {}^7\text{Li} + \nu_e$ | 0.47 (1 ± 0.15) | 0.861 |
| ${}^8\text{B} \rightarrow {}^8\text{Be} + e^+ + \nu_e$ | $5.8 \times 10^{-4}$ (1 ± 0.37) | 15 |
| ${}^{13}\text{N} \rightarrow {}^{13}\text{C} + e^+ + \nu_e$ | 0.06 (1 ± 0.50) | 1.199 |
| ${}^{15}\text{O} \rightarrow {}^{15}\text{N} + e^+ + \nu_e$ | 0.05 (1 ± 0.58) | 1.732 |

The first experiment to detect solar neutrinos was developed by R. Davis. It was based on neutrino capture in the reaction

$$\nu_e + {}^{37}\text{Cl} \rightarrow {}^{37}\text{Ar} + e^+. \tag{4.39}$$

The chief drawback of this reaction is that only neutrinos with energy above 0.81 MeV can be detected. This high threshold energy implies that neutrinos from the primary proton–proton fusion reaction cannot be detected; as indicated in Table 4.1 these have a maximum energy of 0.420 MeV. Furthermore, the neutrinos from electron capture on ${}^7$Be only just exceed the threshold and the probability of capture in ${}^{37}$Cl is exceedingly low. However, most of the neutrinos from ${}^8$B decay have an energy well above the threshold for detection. Indeed, even though these neutrinos contribute a minor component of the neutrino flux from the sun, they are expected to dominate the capture rate in ${}^{37}$Cl. Finally, the neutrinos from the two beta decay processes in the carbon–nitrogen cycle are sufficiently energetic to be detected.

The actual capture rates depend on incident neutrino flux, the number of target ${}^{37}$Cl nuclei and the energy-averaged neutrino capture cross-section. For example, for neutrinos from ${}^8$B decay, the average capture cross-section on ${}^{37}$Cl is $\overline{\sigma} = 1.06 \times 10^{-46}$ m$^2$, and a target containing $N({}^{37}\text{Cl})$ nuclei should give a capture rate of

$$R({}^8\text{B}) = F_\nu({}^8\text{B})\, N({}^{37}\text{Cl})\, \overline{\sigma} = 6.1 \times 10^{-36}\, N({}^{37}\text{Cl}) \text{ per second} \tag{4.40}$$

Because of the low probability of neutrino capture, a special unit called the SNU, or the solar neutrino unit, is used in neutrino astrophysics. This is the capture rate per second per $10^{36}$ target nuclei. We see from Eq. (4.40) that the capture rate of neutrinos from ${}^8$B decay in the sun should be 6.1 SNU. Bahcall also shows that the capture rates of neutrinos from ${}^7$Be, ${}^{13}$N, and ${}^{15}$O are expected to be 1.1 SNU, 0.1 SNU and 0.3 SNU, respectively. In addition, a capture rate of 0.2 SNU is expected from solar neutrinos produced by $p + e^- + p \rightarrow d + \nu_e$. In total the predicted rate for capturing solar neutrinos in ${}^{37}$Cl is

$$\text{Predicted rate} = (7.9 \pm 2.6) \text{ SNU}. \tag{4.41}$$

Davis and his colleagues have taken data and developed their experiment to detect solar neutrinos for over 20 years. It was one of the most ambitious and impressive astrophysics experiments ever embarked upon. The low probability of neutrino capture implies that a huge target containing $^{37}$Cl is needed. This was provided by 610 tons of a dry-cleaning fluid called perchloroethylene, $C_2Cl_4$. By noting that 24% of naturally occurring chlorine is $^{37}$Cl and 76% is $^{35}$Cl, it is easy to show that this target contains $2 \times 10^{30}$ $^{37}$Cl nuclei. A capture rate of 1 SNU in this target yields $2 \times 10^{-6}$ captures a second, or one capture every 6 days. Each capture is only detectable by virtue of the radioactive $^{37}$Ar atom produced. These radioactive argon atoms were flushed out with helium and counted by low background proportional counters. Then the background rate due to cosmic rays had to be subtracted. In 1984 Davis and his collaborators reported a $^{37}$Ar production rate of $0.462 \pm 0.04$ atoms per day against a background rate of $0.08 \pm 0.03$ atoms per day. This corresponds to an observed neutrino capture rate of

$$\text{Observed rate} = (2.05 \pm 0.3)\ \text{SNU}. \qquad (4.42)$$

See Bahcall (1989) for further details.

The discrepancy between the observed and predicted capture rates of solar neutrinos in $^{37}$Cl, Eqs. (4.41) and (4.42), has and continues to be a subject of lively and imaginative debate, a debate which is fully explored in the 500 pages of Bahcall's book *Neutrino Astrophysics*.[2]

Solar neutrinos can also be detected by neutrino–electron scattering. Indeed, data on solar neutrinos by this method was first obtained in 1987 by the Kamiokande II detector in Japan. The Kamiokande detector consists of a huge underground tank of water surrounded by photomultiplier counters which observe the Čerenkov radiation emitted by electrons which have been accelerated to speeds close to the velocity of light by an interaction with an energetic neutrino. This method of detection has important advantages over the $^{37}$Cl experiment. It can record the precise time of arrival and it is sensitive to the direction of the incoming neutrinos. Indeed, a clear peak is seen corresponding to neutrinos coming from the direction of the sun. This experiment also observes a solar neutrino flux that is smaller than expected; the observed rate is about one-half the theoretical prediction.

But it is important to emphasize that the Kamiokande experiment and the $^{37}$Cl experiment can only detect energetic neutrinos. Accordingly both experiments mainly record neutrinos from $^8$B decay in branch III of the proton–proton chain, a very minor part of the hydrogen burning process. Experiments sensitive to branches

---

[2] Because this debate can lead to an uncomfortable feeling called confusion, it is useful to point out at least one positive outcome: the $^{37}$Cl experiment clearly confirms that the carbon–nitrogen cycle plays a minor role in the sun. According to the Standard Solar Model, 1.6% of the luminosity is generated by the carbon–nitrogen cycle and the contribution from the associated neutrinos from $^{13}$N and $^{15}$O beta decay to the solar neutrino capture rate is 0.4 SNU. If the carbon–nitrogen cycle were the dominant mode for hydrogen burning in the sun, the expected capture rate would be about 25 SNU, more than ten times the observed capture rate.

I and II of the chain are needed before firm conclusions can be drawn about hydrogen burning in the sun.

Two such experiments, the SAGE and GALLEX collaborations, began taking data in late 1991. These are radiochemical experiments based on the reaction

$$\nu_e + {}^{71}\text{Ga} \rightarrow {}^{71}\text{Ge} + e^- . \tag{4.43}$$

The threshold energy for this reaction is only 0.233 MeV, well below the maximum energy of neutrinos from the primary proton–proton fusion reaction of the chain, Eq. (4.32). Indeed, these neutrinos should provide half the counting rate in the $^{71}$Ga experiments. But, as in the $^{37}$Cl experiment, there is a formidable problem in identifying the radioactive products of neutrino capture, in this case about one atom of $^{71}$Ge per day in a target containing tons of gallium.

The solar neutrino capture rates in $^{71}$Ga measured by the SAGE and GALLEX collaborations in 1992 were, respectively,

$$[58^{+17}_{-24} \pm 14(\text{systematic})\ \text{SNU}] \quad \text{and} \quad [83 \pm 19 \pm 8(\text{systematic})\ \text{SNU}]. \tag{4.44}$$

These experimental results can be compared with two theoretical predictions: first, a minimum capture rate of 80 SNU in gallium based solely on the requirement that the observed solar luminosity is due to nuclear reactions, regardless of the details of the solar model. Second, an expected rate of 132 SNU predicted by the Standard Solar Model; see Bahcall (1989). The large statistical errors in the experimental results prevent any firm conclusions. However, there is an indication that the number of neutrinos arriving from the sun is less than expected. But it will take several years to collect enough data to confirm this indication.

The premise underlying our account of the solar neutrino problem has been that solar neutrinos, once produced, are well behaved. This may not be the case. Indeed, there is a growing theoretical belief that electron neutrinos emitted during hydrogen burning can change their form to muon neutrinos as they propagate through the sun. Such a transformation could account for the low detection rate of neutrinos on earth. This is interesting particle physics but depressing astrophysics. It implies that the astrophysical information carried by neutrinos may be scrambled.

In view of current theoretical and experimental uncertainties, we have no alternative but to leave this section on solar neutrinos unfinished. The solar neutrino debate has focused attention on the physics and chemistry of neutrino detection, on the reliability of the Standard Solar Model, and on the properties of neutrinos. At present, it is not clear if it is an experimental problem, a solar model problem, or a problem with neutrinos.

## 4.3 HELIUM BURNING

Helium burning produces two vitally important chemical elements, namely oxygen and carbon. Indeed, 65% of your body is oxygen and 18% is carbon.

Moreover, 0.85% and 0.39% of the matter in the solar system is composed of oxygen and carbon; only hydrogen and helium are more abundant. Thus, helium burning is an important process. It is also an interesting process.

Hydrogen burning at the centre of a star ceases when most of the hydrogen in the core has been converted into helium. In the absence of nuclear fusion, the core contracts and gravitational energy is converted into thermal energy. About half of this energy escapes from the core and the other half leads to an increase in temperature. The increased temperature promotes hydrogen burning in a shell surrounding the helium core, and, as more helium is produced, the mass of the central core of helium increases. If the mass of the star is large enough, around 0.5 $M_\odot$ or above, the helium core becomes hot and dense enough for helium burning, which normally takes place at temperatures between 1 to $2 \times 10^8$ K and densities between $10^5$ to $10^8$ kg m$^{-3}$.

The end of hydrogen burning at the centre of a star and the subsequent onset of helium burning has a profound effect on the overall structure and on the outward appearance of the star. The increased internal temperature due to the initial contraction of the helium core leads to an increase in pressure and a large expansion of the outer envelope of the star. When helium burning commences, the energy released causes the core to expand and cool, a cooling that causes a partial contraction of the outer envelope. The net effect is to produce a star with a dense core and a large extended outer envelope, a red giant.

Helium burning is hindered by the absence of stable nuclei with mass 5 and mass 8 to act as stepping-stones to the formation of carbon. But the existence of carbon-based units like you and me implies that there must be a sequence of reactions that produces carbon in stars, a sequence which neatly overcomes the bottle-neck due to the absence of stable nuclei with mass 5 and 8. This sequence was first set out in 1952 by Salpeter. In 1954, Hoyle pointed out that the effectiveness of the sequence depended on the existence of a hitherto unknown excited state of carbon-12. There are three stages to the sequence:

1. The production of a small, but transient, population of unstable $^8$Be nuclei via

$$^4\text{He} + ^4\text{He} \rightleftharpoons ^8\text{Be}. \tag{4.45}$$

2. The production of a small, but transient, population of carbon-12 nuclei in an excited state, denoted by $^{12}$C*, via

$$^4\text{He} + ^8\text{Be} \rightleftharpoons ^{12}\text{C}^*. \tag{4.46}$$

3. The decay of a tiny fraction of carbon-12 nuclei in this excited state to the ground state via

$$^{12}\text{C}^* \rightarrow ^{12}\text{C} + \{2\gamma \quad \text{or} \quad (e^+ + e^-)\}. \tag{4.47}$$

The net effect of this sequence is

$$^4\text{He} + ^4\text{He} + ^4\text{He} \rightarrow ^{12}\text{C} \quad [\text{Q} = +7.275 \text{ MeV}]. \tag{4.48}$$

This fusion of three $^4$He nuclei, or three alpha-particles, is called the triple-alpha process. Note that the first two stages of the triple-alpha process involve reactions which create and destroy nuclei. If these reactions are in thermodynamic equilibrium, it is possible to derive a simple and accurate expression for the rate of the triple-alpha process. To do so, we consider each stage of the process in succession.

## Production of $^8$Be

The ground state of $^8$Be is a state with zero angular momentum and positive parity, $J^\pi = 0^+$. It is unstable because it is more massive than than two $^4$He nuclei, with a mass-energy excess of $(m_8 - 2m_4)c^2 = 91.8$ keV. It decays with a mean lifetime of $\tau = 2.6 \times 10^{-16}$ s into two $^4$He nuclei with the release of 91.8 keV,

$$^8\text{Be} \rightarrow ^4\text{He} + ^4\text{He}. \tag{4.49}$$

Conversely, two $^4$He nuclei can fuse to form a $^8$Be in the endothermic reaction which absorbs 91.8 keV,

$$^4\text{He} + ^4\text{He} \rightarrow ^8\text{Be}. \tag{4.50}$$

In fact, the probability of the interaction of two $^4$He nuclei is enhanced if they approach with a relative energy $E$ near to 91.8 keV and with zero angular momentum. The enhanced probability of interaction arises because they can form an intermediate state, a resonance which corresponds to the ground state of $^8$Be.

The formation of unstable $^8$Be nuclei in a hot gas of ionized helium will be favoured if the resonance with energy of 91.8 keV falls within the energy window for the fusion of two $^4$He nuclei. We recall from Section 4.1, that the joint probability for nuclei to borrow an energy $E$ from a gas at temperature $T$ and to penetrate the Coulomb barrier, which keeps them apart, has a maximum at $E = E_0$ with a width $\Delta$; see Fig. 4.3 and Eqs. (4.23) and (4.25). This energy window for fusion is determined by the Gamow energy for two $^4$He nuclei, which according to Eq. (4.10) is 31.6 MeV, and by the temperature of the gas. A simple calculation shows that the window is in the right place for the formation $^8$Be when the temperature is just above $10^8$ K; in fact, at $T = 1 \times 10^8$ K, the fusion window is just below the resonance at $E = (83 \pm 31)$ keV, whereas at $T = 2 \times 10^8$ K the window is just above at $E = (132 \pm 55)$ keV.

Of course, any $^8$Be nucleus formed will rapidly decay back to two $^4$He nuclei. But at high density and when the temperature is above $10^8$ K, the formation rate can be sufficient to generate a significant population of $^8$Be nuclei in a gas of $^4$He nuclei. Furthermore, if the decay and formation processes, Eqs. (4.49) and (4.50), reach thermodynamic equilibrium, the steady state population of $^8$Be can be found by equating the chemical potential of a $^8$Be nucleus to the chemical potential of a pair of $^4$He nuclei.

According to Eq. (2.21), the chemical potential for nuclei with mass $m_A$ and concentration $n_A$ in a classical gas at temperature $T$ is

$$\mu_A = m_A c^2 - kT \ln\left[\frac{g_A n_{QA}}{n_A}\right], \tag{4.51}$$

where the quantum concentration $n_{QA}$ is given by Eq. (2.22),

$$n_{QA} = \left[\frac{2\pi m_A kT}{h^2}\right]^{3/2}, \tag{4.52}$$

and $g_A$ is an angular momentum multiplicity factor which equals unity for states with zero angular momentum, like the ground states of $^4$He and $^8$Be. Substitution into the equilibrium condition

$$\mu_8 = \mu_4 + \mu_4, \tag{4.53}$$

gives the following result for the population of $^8$Be nuclei in a gas of $^4$He nuclei at temperature $T$:

$$\frac{n_8}{n_4^2} = 2^{3/2}\left[\frac{h^2}{2\pi m_4 kT}\right]^{3/2} \exp[-(m_8 - 2m_4)c^2/kT]. \tag{4.54}$$

Note the key role played by the Boltzmann factor involving the mass-energy difference of $(m_8 - 2m_4)c^2 = 91.8$ keV; this is the energy that must be borrowed from the thermal environment in order to form a $^8$Be.

We can now estimate the population of $^8$Be in a dense, hot gas of helium. For example, if the density of the helium gas is $\rho = 10^8$ kg m$^{-3}$ and the temperature is $T = 2 \times 10^8$ K, the concentration of $^4$He nuclei is $n_4 \approx \rho/m_4 = 1.5 \times 10^{34}$ m$^{-3}$ and the concentration of $^8$Be nuclei is $7 \times 10^{26}$ m$^{-3}$. In other words, there is one $^8$Be nucleus present for every 20 million $^4$He nuclei. However, the Boltzmann factor in Eq. (4.54) indicates that the population of $^8$Be falls off rapidly if the temperature is reduced. It equals $\exp(-5)$ at $2 \times 10^8$ K and $\exp(-10)$ at $1 \times 10^8$ K. At this lower temperature, there is only one $^8$Be for every 2 billion $^4$He nuclei.

We conclude that when the helium core of a star reaches a temperature above $10^8$ K, a tiny fraction of the core is in the form of $^8$Be nuclei in a state of dynamic equilibrium. The turnover of this population is very rapid, with each $^8$Be nucleus existing for an average of $2.6 \times 10^{-16}$ s. Nevertheless, these nuclei provide an adequate raw material for the next stage of the triple-alpha process.

**Production of $^{12}C^*$**

The next stage of the triple-alpha process depends on the existence of a particular excited state of carbon-12, $^{12}C^*$. This seemingly accidental state is so important that its existence was predicted by Hoyle in order to account for helium burning in red giants. In particular, he showed that for helium burning to take place at temperatures as low as $T = 1.2 \times 10^8$ K, there must be a resonant enhancement of the fusion of $^4He$ and $^8Be$. Moreover, he showed that this $^4He$–$^8Be$ resonance had to be at an energy about 300 keV above the threshold. Such a resonance corresponds to an excited state of carbon-12 with an excitation energy of 7.65 MeV.

This excited state of carbon-12 was subsequently found almost exactly where predicted. It has zero angular momentum and even parity, $J^\pi = 0^+$, and the excitation energy above the ground state of carbon-12 is

$$(m_{12}^* - m_{12})c^2 = (7.6542 \pm 0.0015)\ \text{MeV}. \tag{4.55}$$

As illustrated in Fig. 4.6, this state has an energy which is just above the threshold for a $^4He$ and a $^8Be$ nucleus and the threshold for three $^4He$ nuclei. In fact,

$$(m_{12}^* - m_4 - m_8)c^2 = 287.7\ \text{keV} \quad \text{and} \quad (m_{12}^* - 3m_4)c^2 = 379.5\ \text{keV}. \tag{4.56}$$

We have already seen that in a dense gas of helium at a temperature near to $10^8$ K or above, $^4He$ nuclei occasionally fuse to form unstable $^8Be$ nuclei, each of which will usually decay back to two $^4He$ nuclei. We now see that these $^8Be$ nuclei could, very occasionally, fuse with $^4He$ to form $^{12}C^*$, carbon-12 nuclei in the $0^+$ excited state, and that each of these $^{12}C^*$ nuclei would have a brief existence before decaying back to a $^4He$ and a $^8Be$. This will happen if the resonance, which is at an energy of 287.7 keV above the $^4He$–$^8Be$ threshold, is close to the window for the fusion of these two nuclei. It is easy to use Eqs. (4.23) and (4.25) to show that this is the case when the temperature of the gas is just above $10^8$ K; for example, when the temperature is $T = 2 \times 10^8$ K, the window for $^4He$–$^8Be$ fusion is at $E = (232 \pm 73)$ keV.

Thus, when the temperature of an ionized gas of helium exceeds $10^8$ K, collisions between $^4He$ generate small numbers of unstable $^8Be$ and $^{12}C^*$ nuclei. The equilibrium population of $^8Be$ is given by Eq. (4.54). The equilibrium population of $^{12}C^*$ can be found by considering the reactions

$$^4\text{He} + ^8\text{Be} \rightleftharpoons ^{12}\text{C}^*, \tag{4.57}$$

and imposing the condition that

$$\mu_4 + \mu_8 = \mu_{12}^*. \tag{4.58}$$

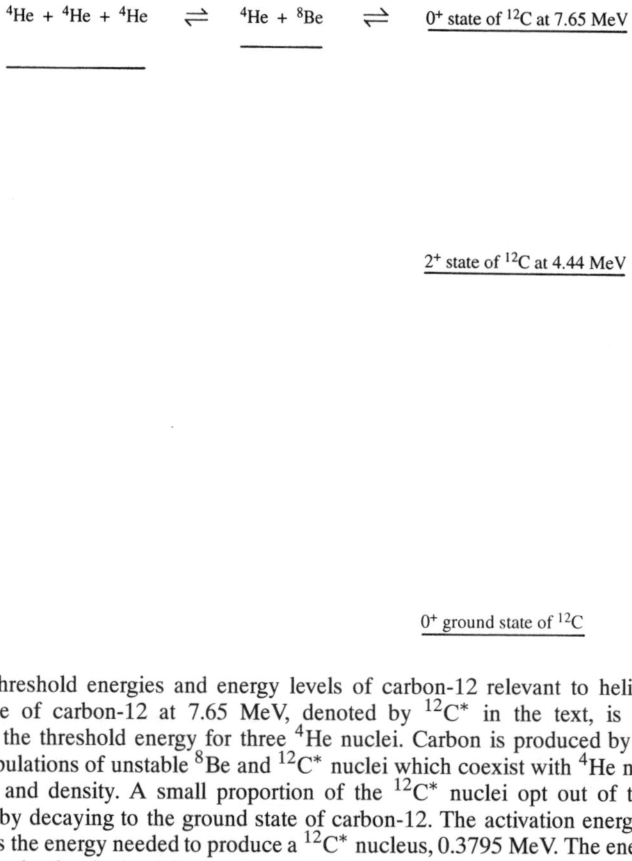

$^4He + {^4He} + {^4He} \rightleftharpoons {^4He} + {^8Be} \rightleftharpoons$ $0^+$ state of $^{12}C$ at 7.65 MeV

$2^+$ state of $^{12}C$ at 4.44 MeV

$0^+$ ground state of $^{12}C$

Fig. 4.6 Threshold energies and energy levels of carbon-12 relevant to helium burning. The $0^+$ state of carbon-12 at 7.65 MeV, denoted by $^{12}C^*$ in the text, is only 0.3795 MeV above the threshold energy for three $^4He$ nuclei. Carbon is produced by establishing transient populations of unstable $^8Be$ and $^{12}C^*$ nuclei which coexist with $^4He$ nuclei at high temperature and density. A small proportion of the $^{12}C^*$ nuclei opt out of this dynamic coexistence by decaying to the ground state of carbon-12. The activation energy for carbon production is the energy needed to produce a $^{12}C^*$ nucleus, 0.3795 MeV. The energy released by carbon production is the difference in energy between the threshold for three $^4He$ nuclei and the ground state of carbon-12, 7.275 MeV.

In complete analogy with Eqs. (4.53) and (4.54), we find that

$$\frac{n_{12}^*}{n_4 n_8} = \left(\frac{3}{2}\right)^{3/2} \left[\frac{h^2}{2\pi m_4 kT}\right]^{3/2} \exp[-(m_{12}^* - m_4 - m_8)c^2/kT]. \tag{4.59}$$

We also recall that the concentration of $^8Be$ nuclei, $n_8$, is given by Eq. (4.54). Thus, we combine Eq. (4.59) with Eq. (4.54), to give

$$\frac{n_{12}^*}{n_4^3} = 3^{3/2} \left[\frac{h^2}{2\pi m_4 kT}\right]^3 \exp[-(m_{12}^* - 3m_4)c^2/kT]. \tag{4.60}$$

Note that, even though the unstable nucleus $^8$Be plays a crucial role in establishing equilibrium, the population of $^{12}$C$^*$ is determined solely by the temperature, the concentration of $^4$He nuclei and the energy difference between a $^{12}$C$^*$ and three $^4$He nuclei. Indeed, Eq. (4.60) can be derived directly by considering the equilibrium established by the reactions

$$^4\text{He} + {}^4\text{He} + {}^4\text{He} \rightleftharpoons {}^{12}\text{C}^*. \tag{4.61}$$

We have already seen that the concentrations of $^4$He and $^8$Be nuclei are $n_4 = 1.5 \times 10^{34}$ m$^{-3}$ and $n_8 = 7 \times 10^{26}$ m$^{-3}$, respectively, in a helium gas with density of $10^8$ kg m$^{-3}$ and temperature $2 \times 10^8$ K. We can now use Eq. (4.60) to show that the concentration of $^{12}$C$^*$ nuclei in such a gas is $n_{12}^* = 3 \times 10^{14}$ m$^{-3}$. We also note that the Boltzmann factor, $\exp[-(m_{12}^* - 3m_4)c^2/kT]$, in Eq. (4.60) implies that the concentration of $^{12}$C$^*$ nuclei falls off rapidly if the temperature is reduced.

Thus, when the helium core of a star reaches a temperature above $10^8$ K, the core contains a small population of $^{12}$C$^*$ nuclei which coexists in dynamic equilibrium with a larger population $^8$Be nuclei and a much larger population of $^4$He nuclei. In the final stage of the triple-alpha process a small fraction of these $^{12}$C$^*$ nuclei opt out from this dynamic coexistence.

### Carbon production

The first two stages of the triple-alpha process create $^8$Be and $^{12}$C$^*$ nuclei via the reactions,

$$^4\text{He} + {}^4\text{He} + {}^4\text{He} \rightleftharpoons {}^4\text{He} + {}^8\text{Be} \rightleftharpoons {}^{12}\text{C}^*. \tag{4.62}$$

Nearly all of the $^{12}$C$^*$ nuclei produced return from whence they came. But a few of them leak away and decay to the ground state of carbon-12 in the following way

$$^{12}\text{C}^* \rightarrow {}^{12}\text{C} + \{2\gamma \quad \text{or} \quad (e^+ + e^-)\}. \tag{4.63}$$

The mean time for this decay is $\tau(^{12}\text{C}^* \rightarrow {}^{12}\text{C}) = 1.8 \times 10^{-16}$ s and the energy released is 7.65 MeV. This irreversible leakage hardly effects the dynamic equilibrium set up by Eq. (4.62), because only a few of the $^{12}$C$^*$ nuclei, roughly one in 2500, opt out by decaying to the ground state. Hence we can still use Eqs. (4.54) and (4.60) to find the populations of $^8$Be and $^{12}$C$^*$ nuclei. The rate of production of carbon-12 nuclei in the ground state is simply the concentration of $^{12}$C$^*$ times the rate at which each of these nuclei opts out; i.e.

$$\frac{dn_{12}}{dt} = \frac{n_{12}^*}{\tau(^{12}\text{C}^* \rightarrow {}^{12}\text{C})}. \tag{4.64}$$

If we use Eq. (4.60), we obtain a production rate of

$$\frac{dn_{12}}{dt} = \frac{n_4^3}{\tau(^{12}C^* \rightarrow ^{12}C)} 3^{3/2} \left[ \frac{h^2}{2\pi m_4 kT} \right]^3 \exp[-(m_{12}^* - 3m_4)\, c^2/kT]. \qquad (4.65)$$

Thus, the carbon production rate in a gas of helium at high temperature is given by the remarkably simple expression, Eq. (4.65). The simplicity arises because of the role played by $^8$Be and $^{12}$C* in establishing thermodynamic equilibrium. We note that the production rate depends on two parameters. The first is the mass-energy difference $(m_{12}^* - 3m_4)c^2$ which enters as an *activation energy* for the process; i.e. as the energy that has to be borrowed to create the key intermediate state, $^{12}$C*. The second is the mean time for $^{12}$C* to decay to the ground state, $\tau(^{12}C^* \rightarrow ^{12}C)$. Both of these parameters have been accurately measured: the activation energy is 0.3795 keV and the mean decay time is $1.8 \times 10^{-16}$ s.

The energy released by the triple-alpha process follows directly from the carbon production rate. To produce a $^{12}$C nucleus, an energy equal to $(m_{12}^* - 3m_4)c^2$ is first absorbed in order to create an intermediate $^{12}$C*, and then an energy equal to $(m_{12}^* - m_{12})c^2$ is released when this intermediate state decays to the ground state. The net energy released is $(3m_4 - m_{12})c^2 = 7.275$ MeV, and the energy production rate by the triple-alpha process is

$$\varepsilon_{3\alpha} = (3m_4 - m_{12})c^2 \frac{dn_{12}}{dt}. \qquad (4.66)$$

As a specific numerical example we reconsider helium burning in helium gas at $T = 2 \times 10^8$ K and $\rho = 10^8$ kg m$^{-3}$. According to Eq. (4.60), $n_{12}^* = 3 \times 10^{14}$ m$^{-3}$. According to Eq. (4.65), carbon-12 nuclei are produced at a rate equal to $1.9 \times 10^{30}$ m$^{-3}$ s$^{-1}$. And according to Eq. (4.66) energy is produced at a rate of $2.2 \times 10^{18}$ W per cubic metre. However, this rate is very sensitive to the temperature. The temperature dependence of the triple-alpha rate is largely governed by the Boltzmann factor in Eq. (4.65). The energy in the exponent, $(m_{12}^* - 3m_4)\, c^2 = 379.5$ keV, is the energy needed to form an intermediate state of a $^{12}$C*; it is the activation energy for carbon production. Because this activation energy is large compared with $kT$ at helium burning temperatures, the Boltzmann factor changes markedly if the temperature is varied. For example, it falls from exp(−22) to exp(−44) when the temperature is reduced from 2 to $1 \times 10^8$ K.

**Carbon consumption**

Once carbon is present at the centre of a red giant, oxygen can be produced by the reaction

$$^4\text{He} + ^{12}\text{C} \rightarrow ^{16}\text{O} + \gamma. \qquad (4.67)$$

The rate of oxygen production can be calculated using the standard equations for thermonuclear fusion outlined in Section 4.1. According to Eq. (4.27), the fusion rate depends largely on the value of the nuclear $S$ factor at an energy $E_0$ in the window for the fusion of $^4$He and $^{12}$C. There are no resonances within or near to this fusion window and the appropriate value of the $S$ factor is small but uncertain; for $E_0$ near to 300 keV, $S$ is in the region of 0.3 MeV barns.

The production of oxygen can be followed by the production of neon via the reaction

$$^4\text{He} + ^{16}\text{O} \rightarrow ^{20}\text{Ne} + \gamma. \tag{4.68}$$

In practice this happens to a minor extent during the helium burning phase. Reaction (4.68) is clearly hindered by an increased Coulomb barrier. Moreover, it is not resonant at energies near to the fusion window. In fact, there is a $J^\pi = 2^-$ excited state of $^{20}$Ne with excitation energy 4.97 MeV, which at first sight could give rise to an enhanced $^4$He–$^{16}$O fusion rate. It does not because a $^4$He and a $^{16}$O cannot couple to form a negative parity state with angular momentum 2; the ground states of both nuclei are $0^+$, and if they couple with relative orbital angular momentum $l = 2$ to give a state with total angular momentum $J = 2$, the parity is positive because $\pi = (-1)^l = +1$. As far as $^4$He and $^{16}$O are concerned, the $2^-$ state of $^{20}$Ne is a state of unnatural parity.

Thus, helium burning dominantly consists of two processes, the triple-alpha process and the production of oxygen by radiative capture of $^4$He by $^{12}$C. The production of heavier nuclei, $^{20}$Ne, $^{24}$Mg, $^{28}$Si, etc by radiative capture of $^4$He does not happen to any appreciable extent during helium burning.

We also note that helium burning involves a jump from helium to carbon. It bypasses the stable nuclei between $A = 6$ and $A = 11$, namely $^6$Li, $^9$Be, $^{10}$B and $^{11}$B. This is consistent with the observed low abundance of these light nuclei in the solar system. These nuclei are not produced in stars, but are primarily produced by spallation reactions in the inter-stellar medium; i.e. collisions between high energy cosmic ray protons and nuclei like $^{12}$C. They were also produced in very small quantities during the big bang.

**What if ?**

We conclude by noting that the outcome of helium burning is finely balanced. During helium burning there exist two competing processes, the carbon producing triple-alpha process, Eq. (4.48), and the carbon consuming oxygen production process, Eq. (4.67). The relative proportion of carbon to oxygen in the solar system is largely an outcome of the balance struck between these two competing processes, and the total amount of carbon and oxygen present depends crucially on the ineffectiveness of the neon production reaction (4.68). It is interesting to speculate on how our surroundings would be affected if this balance was different.

For example, if oxygen production were enhanced by the presence of a resonance near the fusion window, carbon formed by the triple-alpha process would be consumed almost as fast as it is produced. This would lead to an oxygen-rich, carbon-poor environment. A similar situation would arise if the $0^+$ excited state in carbon-12 were a little higher. The triple-alpha rate would be slower, as shown by the Boltzmann factor in Eq. (4.65), and any carbon produced would be rapidly converted into oxygen. Furthermore, if the 4.97 MeV excited state of $^{20}$Ne had quantum numbers $2^+$ instead of $2^-$, most of the carbon and oxygen produced by helium burning would be transformed into neon.

These *what-ifs* are of interest to advocates of the *anthropic principle* which in effect says that physics has to be *just right* in order for biological evolution to be successful. Helium burning seems to fit the principle quite well. It suggests that small changes in seemingly boring excited states of nuclei could easily have led to a solar system in which boredom would not be a problem, because nobody could be around to be bored.

## 4.4 ADVANCED BURNING

As a massive star evolves there is a sequence of nuclear burning stages as the temperature and density at the centre of the star progressively increase. If the mass of a star is large enough, greater than 8 $M_\odot$ or thereabouts, it will evolve beyond helium burning to advanced burning stages involving heavy nuclei. The following processes are thought to occur:

1. When helium burning at the centre of a star ceases, a core of carbon and oxygen contracts and the temperature rises. Carbon burning begins when the temperature approaches $5 \times 10^8$ K at a density of about $3 \times 10^9$ kg m$^{-3}$. Carbon burning produces neon, sodium and magnesium via reactions of the form:

$$^{12}\text{C} + {}^{12}\text{C} \rightarrow {}^{20}\text{Ne} + {}^{4}\text{He}, \tag{4.69}$$

$$^{12}\text{C} + {}^{12}\text{C} \rightarrow {}^{23}\text{Na} + p, \tag{4.70}$$

$$^{12}\text{C} + {}^{12}\text{C} \rightarrow {}^{23}\text{Mg} + n. \tag{4.71}$$

2. Neon burning occurs after carbon burning if the temperature reaches $10^9$ K. At this temperature high-energy thermal photons begin to break up $^{20}$Ne by the photodisintegration reaction

$$\gamma + {}^{20}\text{Ne} \rightarrow {}^{16}\text{O} + {}^{4}\text{He}. \tag{4.72}$$

The $^4$He nuclei released can then react with undissociated $^{20}$Ne nuclei to form $^{24}$Mg,

$$^{4}\text{He} + ^{20}\text{Ne} \rightarrow ^{24}\text{Mg} + \gamma. \tag{4.73}$$

3. After neon burning, the core of the star consists mainly of $^{16}$O and $^{24}$Mg. The oxygen burning phase begins if the temperature reaches $2 \times 10^9$ K, the most important product being $^{28}$Si which is produced by the reaction

$$^{16}\text{O} + ^{16}\text{O} \rightarrow ^{28}\text{Si} + ^{4}\text{He}. \tag{4.74}$$

4. Silicon burning begins if the temperature reaches $3 \times 10^9$ K. At this temperature silicon is gradually destroyed by high energy thermal photons releasing $^4$He nuclei, protons and neutrons. These light particles then combine with undissociated nuclei to build more massive nuclei. A complex network of capture and photodisintegration reactions compete with each other, and the net effect is that loosely bound nuclei tend to be transformed into nuclei of higher stability.

To some extent these advanced burning phases involve the same physics of thermonuclear fusion that we encountered when we considered hydrogen and helium burning. But a new type of physical phenomenon occurs when the temperature rises above $10^9$ K: nuclei are broken up by high energy thermal photons and the nuclear material so formed is then reduced to its most stable form. We shall briefly illustrate these ideas by considering the role of the photodisintegration of nuclei in silicon burning.

The photodisintegration of nuclei is the nuclear physics analogue of the ionization of atoms. We recall from Section 2.5, that atomic ionization becomes important at about 3000 K. Since nuclear binding energies are typically a million times larger than atomic binding energies, nuclear photodisintegration becomes appreciable at a temperature which is about a million times higher than 3000 K, i.e. about $3 \times 10^9$ K.

As we have mentioned, a network of competing photodisintegration and capture reactions occurs during the silicon burning stage. To illustrate the underlying principle we consider part of this network which is initiated by the production of $^4$He nuclei by the photodisintegration of the tightly bound nucleus $^{28}$Si,

$$\gamma + ^{28}\text{Si} \rightarrow ^{24}\text{Mg} + ^{4}\text{He}. \tag{4.75}$$

This process proceeds slowly because the thermal photon must have an energy above 9.98 MeV. The $^4$He nuclei released by this photodisintegration can induce a sequence of reactions which produces sulphur, argon, calcium, . . ., etc, as follows

$$^{28}\text{Si} + ^{4}\text{He} \rightleftharpoons ^{32}\text{S} + \gamma,$$

$$^{32}\text{S} + ^{4}\text{He} \rightleftharpoons ^{36}\text{Ar} + \gamma,$$

$$^{36}\text{Ar} + ^{4}\text{He} \rightleftharpoons ^{40}\text{Ca} + \gamma,$$

and so on until

$$^{52}\text{Fe} + ^{4}\text{He} \rightleftharpoons ^{56}\text{Ni} + \gamma.$$

These reactions can take place more rapidly than the initial photodisintegration of $^{28}$Si which initiates this build-up process by releasing $^{4}$He nuclei. In fact, the time scale for the build-up of heavier nuclei is governed by the slow photodisintegration of the tightly bound $^{28}$Si. Because the build-up reactions approach thermodynamic equilibrium, relative concentrations can be found by equating chemical potentials. For example, if we consider the first reaction in the sequence and equate the sum of the chemical potentials of $^{28}$Si and $^{4}$He to the chemical potential of $^{32}$S,

$$\mu_{28} + \mu_4 = \mu_{32}, \tag{4.76}$$

we find that the concentrations of these nuclei are related by

$$\frac{n_{28}n_4}{n_{32}} \approx \left[\frac{2\pi m_r kT}{h^2}\right]^{3/2} \exp[-Q/kT], \tag{4.77}$$

where $Q$ is the energy needed to release a $^{4}$He from a $^{32}$S nucleus,

$$Q = [m_{28} + m_4 - m_{32}]c^2 \approx 6.95 \text{ MeV},$$

and $m_r$ is the reduced mass for the $^{28}$Si$-^{4}$He system, 3.5 amu. We note from Eq. (4.77) that the abundance of $^{32}$S relative to $^{28}$Si is determined by the temperature and the concentration of $^{4}$He during silicon burning. If this concentration is $10^{34}$ m$^{-3}$, say, then the equilibrium ratio of $^{32}$S to $^{28}$Si is about 1 to 4 at a temperature of $5 \times 10^9$ K.

Similar considerations can be used to explore the abundances of $^{36}$Ar, $^{40}$Ca, ... etc relative to $^{28}$Si. We note that the abundances of these nuclei, and others, will be governed by Boltzmann factors, $\exp[-Q/kT]$, involving their break-up energy. As a result there will always be a tendency for the more tightly bound nuclei to be favoured.

We recall from Fig. 1.3 that the binding energy per nucleon increases as the mass number $A$ approaches 56. Hence the breakup of $^{28}$Si and the subsequent rearrangement of the nucleons favours the formation of the most stable nuclei in the periodic table near $A$ = 56, namely the isotopes of Cr, Mn, Fe, Co and Ni. Because the binding energy per nucleon reaches a maximum at $A$ = 56, energy is absorbed from the gas if light particles are captured to form nuclei with $A > 56$. For this reason nuclei beyond the iron group in the periodic table are not formed during silicon burning.

Thus, silicon burning is a sequence of radiative capture and photodisintegration reactions which in effect *melt* silicon in a sea of alpha particles, protons and neutrons to create heavier elements with mass number in the range 30 to 56. Indeed, silicon burning is often referred to as silicon melting. This terminology has the advantage of stressing that the underlying mechanism is different. Apart from neon burning, it is the only stage of nuclear burning that involves the disintegration of nuclei by high energy thermal photons.

Finally, we point out that the time scales involved in the advanced burning stages are much shorter than the time scale for hydrogen or helium burning. The primary reason is that the rate of production of nuclear energy is governed by the rate of energy loss from the star. The energy loss is large at the high temperatures reached during the advanced burning stages. For example, at these temperatures neutrinos can be produced by electron–positron collisions and by other mechanisms, and energy loss by neutrino emission can be large. The nuclear burning time scales for a star of mass 25 $M_{\odot}$ are listed in Table 4.2. The nuclear burning sequence terminates with silicon burning which yields a central core composed of the most stable nuclei in the periodic table with mass number near 56, from which no further energy can be extracted. We shall see in Chapter 6, that a star which evolves beyond silicon burning is heading for some sort of catastrophe because the central core will collapse under gravity when its mass exceeds the Chandrasekhar limit of about 1.4 $\mathrm{M}_{\odot}$.

TABLE 4.2 The time scale for the nuclear burning stages for a star of mass 25 $M_{\odot}$, and the central temperature and density at which they take place. This data is based on the calculations of Weaver, cited by Rolfs and Rodney (1988).

| Stage | Time scale | Temperature ($10^9$ K) | Density (kg $m^{-3}$) |
|---|---|---|---|
| Hydrogen burning | $7 \times 10^6$ years | 0.06 | $5 \times 10^4$ |
| Helium burning | $5 \times 10^5$ years | 0.23 | $7 \times 10^5$ |
| Carbon burning | 600 years | 0.93 | $2 \times 10^8$ |
| Neon burning | 1 year | 1.7 | $4 \times 10^9$ |
| Oxygen burning | 6 months | 2.3 | $1 \times 10^{10}$ |
| Silicon burning | 1 day | 4.1 | $3 \times 10^{10}$ |

## SUMMARY

### The physics of nuclear fusion

- Nuclear fusion can take place at an energy $E$ below the Coulomb barrier which tends to keep the nuclei apart. The probability of penetrating this barrier is approximately given by

$$\text{Probability of penetration} \approx \exp\left[-\left(\frac{E_G}{E}\right)^{1/2}\right], \tag{4.12}$$

where $E_G$, the Gamow energy, depends on the electric charge on the nuclei, $Z_A$ and $Z_B$, and their reduced mass, $m_r$. It is given by

$$E_G = (\pi\alpha Z_A Z_B)^2 2m_r c^2. \tag{4.10}$$

- The energy dependence of the fusion cross-section, $\sigma(E)$, is invariably dominated by the rapidly rising probability of penetration of the Coulomb barrier. It is usually represented by

$$\sigma(E) = \frac{S(E)}{E}\exp\left[-\left(\frac{E_G}{E}\right)^{1/2}\right]. \tag{4.16}$$

The function $S(E)$ is determined by the nuclear physics of fusion and varies slowly with energy except at energies near to nuclear resonances.

- The thermonuclear reaction rate per unit volume for the fusion of two nuclei, $A$ and $B$, with concentrations $n_A$ and $n_B$ is given by

$$R_{AB} = n_A n_B \langle \sigma v_r \rangle. \tag{4.19}$$

If the nuclei form a classical, non-relativistic gas at a temperature $T$, their relative velocity $v_r$ has a Maxwellian distribution, Eq. (4.20), and the fusion rate is given by Eq. (4.22)

$$R_{AB} = n_A n_B \left[\frac{8}{\pi m_r}\right]^{1/2}\left[\frac{1}{kT}\right]^{3/2}\int_0^\infty S(E)\exp\left[-\frac{E}{kT} - \left(\frac{E_G}{E}\right)^{1/2}\right] \mathrm{d}E. \tag{4.22}$$

- The exponential in this equation is proportional to the joint probability that an energy $E$ is borrowed from the thermal environment and that the Coulomb barrier is penetrated. This joint probability has a peak at $E = E_0$ and a width $\Delta$ as shown in Fig. 4.3, where according to Eqs. (4.23) and (4.25)

$$E_0 = \left[\frac{E_G (kT)^2}{4}\right]^{1/3} \quad \text{and} \quad \Delta = \frac{4}{3^{1/2}2^{1/3}} E_G^{1/6}(kT)^{5/6}.$$

As a result thermonuclear fusion predominantly takes place at energies in the energy window $E_0 \pm \Delta/2$.

- If the nuclear fusion factor $S(E)$ is approximately constant for energies in the fusion window, the fusion rate is given by

$$R_{AB} = 6.48 \times 10^{-24}\frac{n_A n_B}{A_r Z_A Z_B} S(E_0)\left(\frac{E_G}{4kT}\right)^{2/3}\exp\left[-3\left(\frac{E_G}{4kT}\right)^{1/3}\right] \mathrm{m^{-3}s^{-1}}. \tag{4.27}$$

This expression implies that the temperature dependence of the fusion rate is approximately given by

$$R_{AB} \propto T^a \quad \text{where} \quad a \approx \left[\frac{E_G}{4kT}\right]^{1/3}.$$

**Hydrogen burning**

- Hydrogen burning transforms protons into $^4$He nuclei. The most likely transformation involves the emission of two positrons and two electron neutrinos, i.e.

$$4p \rightarrow {}^4\text{He} + 2e^+ + 2\nu_e.$$

When the annihilation energy of the positrons is included, the energy release is 26.73 MeV per $^4$He formed; a small percentage of this energy is carried by the neutrinos which escape almost without interaction.
- In the sun, hydrogen burning occurs mostly by the proton–proton chain illustrated in Fig. 4.4. There are three main branches denoted by I, II, and III. The overall rate of the proton–proton chain is governed by the first reaction in the chain,

$$p + p \rightarrow d + e^+ + \nu_e. \tag{4.32}$$

In the sun the proton–proton chain is terminated by branch I 85% of the time, by branch II 15% of the time and by branch III 0.02% of the time. The energy released by the chain is 15 MeV per proton–proton fusion. This energy release together with the calculated proton–proton fusion rate leads to an energy production rate by the proton–proton chain given by

$$\varepsilon_{pp} = 9.5 \times 10^{-37} X_1^2 \rho^2 T^4 \quad \text{W m}^{-3}, \tag{4.37}$$

where $X_1$ is the mass fraction of hydrogen.
- In main sequence stars more massive than the sun, hydrogen burning predominantly takes place via the carbon–nitrogen cycle illustrated in Fig. 4.5. This is a cycle of reactions in which $^{12}$C acts as a catalyst in the transformation of protons to $^4$He nuclei.
- In principle, neutrinos released during hydrogen burning could provide direct information on the actual reactions involved. The expected flux of neutrinos from the sun due to particular reactions in the proton–proton chain and in the carbon–nitrogen cycle are listed in Table 4.1. Experiments designed to detect solar neutrinos with an energy above 0.81 MeV consistently yield measured fluxes below the theoretically expected result; see Eqs. (4.41) and (4.42).

Experiments designed to detect the more numerous, low energy neutrinos may resolve the solar neutrino problem.

## Helium burning

- Helium burning produces carbon, some of which is converted to oxygen.
- Carbon is produced by the triple-alpha process, Eqs. (4.45) to (4.47). The net effect is

$$^4\text{He} + ^4\text{He} + ^4\text{He} \rightarrow ^{12}\text{C} \quad [\text{Q} = +7.725\ \text{MeV}]. \tag{4.48}$$

Unstable $^8$Be nuclei and carbon-12 nuclei in the excited $0^+$ state ($^{12}$C*), play a key role in the triple-alpha process. Collisions between $^4$He nuclei generate small numbers of $^8$Be and $^{12}$C* nuclei via the reactions

$$^4\text{He} + ^4\text{He} + ^4\text{He} \rightleftharpoons ^4\text{He} + ^8\text{Be} \rightleftharpoons ^{12}\text{C}^*.$$

The populations of the $^8$Be and $^{12}$C* nuclei are given by Eq. (4.54) and Eq. (4.60), respectively. A tiny fraction of the $^{12}$C* nuclei leak away from this dynamic coexistence by decaying to the ground state of $^{12}$C in accordance with Eq. (4.63). The production rate of carbon-12 nuclei in the ground state is given by

$$\frac{\mathrm{d}n_{12}}{\mathrm{d}t} = \frac{n_4^3}{\tau(^{12}\text{C}^* \rightarrow ^{12}\text{C})} 3^{3/2} \left[\frac{h^2}{2\pi m_4 kT}\right]^3 \exp[-(m_{12}^* - 3m_4)c^2/kT]. \tag{4.65}$$

This rate is governed by two parameters, the energy needed to form a $^{12}$C*, the activation energy $(m_{12}^* - 3m_4)c^2$, and the mean time for a $^{12}$C* to decay to the ground state, $\tau(^{12}\text{C}^* \rightarrow ^{12}\text{C})$.
- Some of the carbon produced is transformed into oxygen by the reaction

$$^4\text{He} + ^{12}\text{C} \rightarrow ^{16}\text{O} + \gamma. \tag{4.67}$$

- The relative proportion of carbon and oxygen produced by helium burning depends on the relative effectiveness of the carbon-producing triple-alpha process and the oxygen-production reaction which consumes carbon. Nuclei heavier than $^{16}$O are not produced in any quantity during helium burning because the rate of capture of $^4$He by $^{16}$O, Eq. (4.68), is slow.

## Advanced burning

- If the mass of a star exceeds a value of about $8M_\odot$, it will evolve beyond helium burning. Carbon, neon, oxygen and silicon burning stages can occur. See Table 4.2.

- The main new physical phenomenon that arises in advanced burning is the photodisintegration of nuclei by high energy thermal photons when the temperature is above $10^9$ K. Indeed, silicon burning involves a rearrangement of nuclear material by a network of photodisintegration and capture reactions. Because this rearrangement tends to reduce nuclei to their most stable form, nuclei close to iron in the periodic table are formed.
- Once a star evolves beyond silicon burning, no further energy can be extracted from nuclear reactions in its iron core. The core of such a star collapses when its mass exceeds the Chandrasekhar limit of about $1.4 M_\odot$.

## PROBLEMS 4

4.1 Find the classical distance of closest approach for two protons with an energy of approach equal to 2 keV. Estimate the probability that the protons penetrate the Coulomb barrier tending to keep them apart. Compare this probability with the corresponding probability for two $^4$He nuclei with the same energy of approach.

4.2 We have seen that the quantum-mechanical penetration of a Coulomb barrier plays a crucial role in thermonuclear fusion. It also plays a crucial role in the alpha-decay of nuclei such as $^{235}$U. In the simplest model for alpha-decay, the alpha-particle is preformed and trapped within the nucleus by a potential similar to that shown in Fig. 4.1. The mean rate of decay, $\lambda$, is then a frequency $\nu$ with which the alpha-particle hits the confining barrier times a probability of penetration of the Coulomb barrier; this probability is given by Eq. (4.12). Write down an approximate expression for the decay rate in terms of $\nu$, $E_G$ and the energy released by alpha-decay, $E$. The half-life for the alpha-decay of $^{235}$U is $\tau_{1/2} = 0.69/\lambda = 7.1 \times 10^8$ years and the energy released, $E$, is 4.68 MeV. The energy released by the alpha-decay of $^{239}$Pu is 5.24 MeV. Estimate the half-life of this isotope of plutonium.

4.3 Assume that the solar luminosity of $4 \times 10^{26}$ W is due to hydrogen burning by the proton–proton chain illustrated in Fig. 4.4. The expected flux of neutrinos from the primary proton–proton fusion reaction is then almost fixed if the relative importance of branch I and II is known. Find an upper limit and a lower limit for this neutrino flux.

4.4 The flux of energetic neutrinos from $^8$B decay in branch III of the proton–proton chain is very dependent on the central temperature of the sun. Confirm this by showing that the rate of the reaction producing $^8$B,

$$p + {}^7\mathrm{Be} \rightarrow {}^8\mathrm{B} + \gamma,$$

is approximately proportional to $T^{14}$, when the temperature $T$ is near to $1.5 \times 10^7$ K. [In fact, the local production rate of neutrinos from $^8$B decay is found to be proportional to $T^{24}$ when the temperature dependence of the reactions leading to $^7$Be formation is taken into account; see Bahcall (1989).]

4.5 Consider hydrogen burning by the carbon–nitrogen cycle illustrated in Fig. 4.5. Show that, at a temperature of $1.5 \times 10^7$ K, the slowest reaction in the cycle is

$$p + {}^{14}\mathrm{N} \rightarrow {}^{15}\mathrm{O} + \gamma,$$

and thereby estimate the temperature dependence of the carbon–nitrogen cycle in the sun. It is thought that about 1.6% of the solar luminosity is generated by the carbon–nitrogen cycle. Estimate by how much this would change if the central temperature of the sun were increased by 1%.

4.6 Calculate the power per kilogram produced by helium burning in helium when the density is $10^8$ kg m$^{-3}$ and the temperature is $10^8$ K. By how much would this power change if the excitation energy of the $0^+$ state of carbon-12 were 7.66 MeV instead of 7.65 MeV?

4.7 The photodisintegration of nuclei plays an increasingly important role as a star evolves and as the temperature at its centre increases. Use the results of Section 2.3 to obtain an expression for the number of photons per unit volume in a gas at temperature $T$ with an energy above 9.98 MeV, the minimum energy needed to eject a $^4$He nucleus from a $^{28}$Si nucleus. Estimate the fractional change in the number of such photons that occurs when the temperature rises from $1 \times 10^9$ K to $4 \times 10^9$ K.

4.8 The practical exploitation of thermonuclear fusion as a energy source on earth depends on raising the temperature of a plasma containing ionized deuterium and tritium to an ignition temperature $T_{ign}$ of about $2 \times 10^8$ K; at this temperature the rate of energy production by the fusion of deuterons and tritons is faster than the rate at which energy is lost by radiation. A considerable amount of energy is needed to heat the plasma to the ignition temperature. A greater amount of energy must be released by fusion in order to make the process cost effective. Hence, the hot plasma must be confined for a certain minimum time. Show that, if the plasma contains equal numbers of deuterons and tritons, this minimum confinement time, $\tau$, is approximately given by

$$n_i \tau > \frac{12kT_{ign}}{\langle \sigma v_r \rangle Q_{dep}}$$

where $n_i$ is the number of ions per unit volume, $\sigma$ is the deuteron–triton fusion cross-section and $Q_{dep}$ is the energy deposited in the plasma per fusion. (This result is called the Lawson Break-Even Condition. For a deuterium–tritium plasma at $2 \times 10^8$ K, $n_i \tau$ must exceed $1.6 \times 10^{20}$ m$^{-3}$ s. This implies that a dilute plasma with $n_i = 10^{20}$ m$^{-3}$ must be confined for at least 1.6 s to break even.)

CHAPTER 5

# Stellar structure

A complete analysis of stellar structure requires calculations of considerable complexity and the numerical solution of a coupled set of differential equations. The aim of this chapter is very modest by comparison. It is to use simple models to gain insight into some of the most basic ideas of stellar structure. The discussion will be restricted to stars with homogeneous chemical compositions. We shall consider the structure of main sequence stars, like the sun. We shall estimate the minimum mass of a main sequence star by considering the temperature needed for hydrogen burning, and the maximum mass by considering the destabilizing effect of radiation pressure. These lower and upper limits to stellar masses are shown to be comparable with a fundamental stellar mass $M_{\star}$, whose value is determined by the mass of the nucleon and a dimensionless measure of the strength of the gravitational interaction between nucleons. We shall begin by reminding the reader of some concepts from earlier chapters that are particularly relevant to our discussion of stellar structure.

## 5.1 PREAMBLE

During most of its existence a star is in a state which evolves very slowly, a state which is very close to hydrostatic and thermodynamic equilibrium. The internal pressure gradient is just sufficient to hold up the star and according to Eq. (1.5)

$$\frac{\mathrm{d}P}{\mathrm{d}r} = -\frac{Gm(r)\rho(r)}{r^2}, \tag{5.1}$$

where $m(r)$, the mass enclosed by a sphere of radius $r$, is given by

$$\frac{\mathrm{d}m}{\mathrm{d}r} = 4\pi r^2 \rho(r). \tag{5.2}$$

The internal temperature gradient is just sufficient to maintain the power flow towards the surface. If energy transport is by radiative diffusion we can use Eq. (3.28) and write

$$\frac{\mathrm{d}T}{\mathrm{d}r} = -\frac{3}{4ac}\frac{\kappa(r)\rho(r)}{[T(r)]^3}\frac{L(r)}{4\pi r^2} \tag{5.3}$$

where, according to Eq. (3.26),

$$\frac{\mathrm{d}L}{\mathrm{d}r} = 4\pi r^2 \varepsilon(r). \tag{5.4}$$

We recall that $L(r)$ is the power generated within a sphere of radius $r$ and $\varepsilon(r)$ is the power density at $r$.

Equations (5.1) to (5.4) are the fundamental equations of stellar structure. They are based on the assumptions of spherical symmetry, hydrostatic equilibrium under Newtonian gravity and the flow of energy by radiative diffusion. The latter assumption often has to be modified to allow for energy transport by convection, as in the outer layers of the sun or in the core of a massive main sequence star, or energy transport by conduction as in a white dwarf.

The static structure of a star can be found if the fundamental equations of stellar structure are supplemented by equations which relate the pressure, opacity and power to the density and temperature of the stellar material; i.e.

$$P = P(\rho,T), \quad \kappa = \kappa(\rho,T) \quad \text{and} \quad \varepsilon = \varepsilon(\rho,T). \tag{5.5}$$

These equations for the properties of matter and radiation inside the star have been discussed in earlier chapters. We shall now recall some of the most relevant results.

In Chapter 2 we considered the pressure generated by matter and radiation. Three important sources of this pressure were identified, namely classical electrons and ions, degenerate electrons, and radiation or photons. The number densities of electrons and ions in a completely ionized plasma depend on the mass fraction of hydrogen, helium and heavier elements, $X_1$, $X_4$ and $X_A$, in the plasma. According to Eqs. (2.67) and (2.69)

$$n_e \approx [1 + X_1]\rho/2m_{\mathrm{H}}, \tag{5.6}$$

$$n_i \approx [2X_1 + 0.5X_4]\rho/2m_{\mathrm{H}}, \tag{5.7}$$

and

$$n = n_e + n_i \approx [1 + 3X_1 + 0.5X_4]\rho/2m_{\rm H}. \tag{5.8}$$

As discussed in Section 2.1 these particles can form a classical or a quantum gas. In particular, at low density the electrons and ions form an ideal classical gas with a pressure

$$P = n_e kT + n_i kT = nkT. \tag{5.9}$$

At high densities the electrons form a degenerate quantum gas with a pressure given by Eq. (2.31),

$$P = K_{NR} n_e^{5/3}, \quad \text{where} \quad K_{NR} = \frac{h^2}{5m_e}\left[\frac{3}{8\pi}\right]^{2/3}. \tag{5.10}$$

But at very high densities the degenerate electrons become ultra-relativistic and the pressure approaches a value given by Eq. (2.34),

$$P = K_{UR} n_e^{4/3}, \quad \text{where} \quad K_{UR} = \frac{hc}{4}\left[\frac{3}{8\pi}\right]^{1/3}. \tag{5.11}$$

Finally at high temperature, the pressure due to radiation or photons can be comparable with the gas pressure due to electrons and ions. The radiation pressure at a temperature $T$ is given by Eq. (2.44),

$$P = \frac{1}{3}aT^4 \quad \text{with} \quad a = \frac{8\pi^5 k^4}{15h^3c^3}. \tag{5.12}$$

The opacity of the stellar medium is determined by the interaction of radiation with electrons, ions and atoms. Three processes were mentioned in Chapter 3. Thomson scattering by electrons yields a constant background opacity given by Eq. (3.18),

$$\kappa_{es} = (1 + X_1)\ 0.02\ \text{m}^2\ \text{kg}^{-1}. \tag{5.13}$$

Bound–free absorption and free–free absorption give rise to a density and temperature dependent opacity described by Kramers' law, Eq. (3.17),

$$\kappa \propto \rho T^{-3.5}. \tag{5.14}$$

Bound–free absorption and free–free absorption are important at the low temperatures and high densities found in main sequence stars like the sun. But the constant background opacity due to Thomson scattering is dominant at the higher temperatures and lower densities found in main sequence stars more massive than the sun.

The generation of thermonuclear power was discussed in Chapter 4. Several important chains of reactions were identified. In this chapter we shall only be interested in hydrogen burning in main sequence stars. In the sun, hydrogen burning is dominated by the reactions of the proton–proton chain; according to Eq. (4.37), the power generated is given approximately by

$$\varepsilon_{pp} = 9.5 \times 10^{-37} X_1^2 \rho^2 T^4 \text{ W m}^{-3}. \tag{5.15}$$

## 5.2 SIMPLE STELLAR MODELS

The fundamental equations of stellar structure, Eqs. (5.1) to (5.4), reduce the problem of calculating the structure of a star to the solution of four coupled first-order differential equations in four unknown functions $P(r)$, $m(r)$, $T(r)$ and $L(r)$. Clearly four boundary conditions are needed to specify a unique solution. Two of these are straightforward. They are $m(0) = 0$ and $L(0) = 0$, which are satisfied because the mass and the energy generated within a sphere of radius $r$ must tend to zero as $r$ tends to zero. Two other boundary conditions can be obtained by specifying the pressure and the temperature near the surface of the star; this, in practice, requires some knowledge of the properties of the stellar atmosphere. We shall not discuss this procedure further. Instead, we shall consider approximate models of stellar structure based on Eqs. (5.1) to (5.4) which are simple enough to permit physical insight into some of the general features of stellar structure.

A traditional way of proceeding is to combine the equation of hydrostatic equilibrium and the equation describing the conservation of mass, Eqs. (5.1) and (5.2), to give the second-order equation

$$\frac{1}{r^2}\frac{\mathrm{d}}{\mathrm{d}r}\left[\frac{r^2}{\rho}\frac{\mathrm{d}P}{\mathrm{d}r}\right] = -4\pi G\rho. \tag{5.16}$$

This equation involves two unknown functions, $P(r)$ and $\rho(r)$. It can be reduced to an equation in one unknown function by assuming a simple relation between the pressure and the density which is valid throughout the star. This is the procedure adopted in polytrope models for stellar structure. In particular, a polytrope model with index $n$ is obtained by imposing the following relation between pressure and density,

$$P = K\rho^{(n+1)/n}, \tag{5.17}$$

where $K$ is a constant. When this relation is substituted into Eq. (5.16), we obtain the following non-linear second order differential equation for the density inside the star:

$$\frac{1}{r^2}\frac{\mathrm{d}}{\mathrm{d}r}\left[\frac{r^2}{\rho}\frac{\mathrm{d}}{\mathrm{d}r}(K\rho^{(n+1)/n})\right] = -4\pi G\rho. \tag{5.18}$$

A unique numerical solution of this equation can be obtained by imposing two boundary conditions. Two such conditions on the function $\rho(r)$ are

$$\rho = \rho_c \quad \text{and} \quad \frac{\mathrm{d}\rho}{\mathrm{d}r} = 0 \quad \text{at } r = 0, \tag{5.19}$$

where the second condition follows immediately from the substitution of Eq. (5.17) into Eq. (5.1) and the use of the boundary condition $m(0) = 0$. Thus, once a value for the central density $\rho_c$ is fixed, the numerical solution of Eq. (5.18) gives a density profile $\rho(r)$. The radius $R$ of the star is then the value for $r$ for which $\rho(r)$ is zero and the total mass of the star is $M = m(R)$. The pressure inside the star can be found using Eq. (5.17) and the thermal properties of the star can then be deduced using an equation of state $P = P(\rho,T)$ and the stellar equations (5.3) and (5.4).

Polytrope models based upon the simple relation between pressure and density, Eq. (5.17), have played an important role in development of stellar structure theory, particularly before the advent of powerful computers. Accordingly, they are fully described in many books on astrophysics. However, despite the drastic simplification represented by Eq. (5.17), polytrope models still involve a numerical solution of a messy differential equation.

A simpler approach is to guess a suitable form for the density profile within the star, $\rho = \rho(r)$, and to use this as a starting point for an approximate solution of Eqs. (5.1) to (5.4). These equations can then be tackled sequentially. Equations (5.1) and (5.2) can be integrated to give a profile for the pressure $P(r)$. An equation of state $P = P(\rho,T)$ can then be used to find the temperature profile $T(r)$, which can be combined with an opacity $\kappa = \kappa(\rho,T)$ to estimate the power flow $L(r)$ by using Eq. (5.3). This power flow can then be compared with the power flow $L(r)$ found by integrating Eq. (5.4) using a nuclear power density $\varepsilon = \varepsilon(\rho,T)$. There is, of course, no guarantee that the two expressions for $L(r)$ will be similar, and, in practice, there is almost no similarity if the initial guess for the density profile is a simple function.

A related approach was proposed by Clayton in 1986. The starting point is a simple parametrization of the pressure profile $P = P(r)$ within the star. This is more successful than the approach based upon an initial choice for the density $\rho(r)$, because the choice for the pressure can be shaped by constraints directly imposed by hydrostatic equilibrium. We shall see that the Clayton model can yield reasonably correct answers when applied to the sun.

### Pressure inside a star

The pressure at the centre of the star will greatly exceed the average pressure inside the star. For example, the pressure at the centre of the sun is about $2 \times 10^{16}$ Pa, which is about 200 times the average value given by Eq. (1.7), namely

$$\langle P \rangle = -\frac{E_{GR}}{3V}, \tag{5.20}$$

where $E_{GR}$, the gravitational potential energy, is

$$E_{GR} \approx \frac{GM_\odot^2}{R_\odot}. \tag{5.21}$$

In order to model the large variation of the pressure inside a star, we note that the pressure gradient is directly constrained by the equation for hydrostatic equilibrium (5.1). It is easy to show that this equation implies that the pressure gradient tends to zero at the centre and at the surface. Near the centre, where $r$ is small, the enclosed mass $m(r)$ is approximately equal to $4\pi r^3/3$ times the central density $\rho_c$, and Eq. (5.1) becomes

$$\frac{dP}{dr} = -\frac{4\pi}{3} G\rho_c^2 r. \tag{5.22}$$

Near the surface, where $r$ encloses most of the stellar matter, $m(r)$ is approximately equal to $m(R)$ or the total mass $M$, and Eq. (5.1) becomes

$$\frac{dP}{dr} = -\frac{GM\rho(r)}{r^2}. \tag{5.23}$$

Thus, hydrostatic equilibrium demands that the pressure gradient inside a star is zero at the centre, that it initially varies linearly with $r$, but that it eventually approaches zero again when the density decreases near the surface. The essence of the Clayton model is to guess a simple form for the pressure profile inside the star which takes these constraints into account. Such a guess can be a reasonable starting point for a stellar structure calculation, particularly for a star with a homogeneous chemical composition.

If the chemical composition of the star is uniform, the variation in the pressure should be smooth, as shown in Fig. 5.1. Following Clayton, we shall model the pressure gradient inside such a star by the following expression for the pressure gradient:

$$\frac{dP}{dr} = -\frac{4\pi}{3} G\rho_c^2 r \exp(-r^2/a^2), \tag{5.24}$$

where $a$ is a length parameter which is yet to be specified. This expression gives an accurate representation of the pressure gradient at small $r$; see Eq. (5.22). In

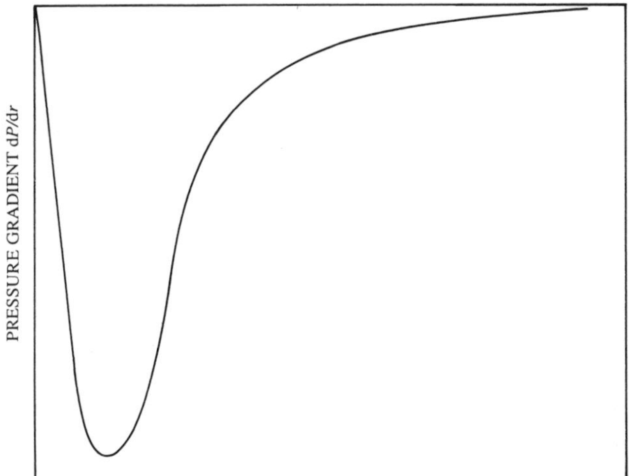


Fig. 5.1 The typical variation in the pressure gradient d$P$/d$r$ inside a star with a homogeneous chemical composition. Near the centre the pressure gradient varies linearly with $r$, and near the surface it is proportional to $\rho(r)/r^2$.

contrast, the representation at large $r$ is very approximate. However, the necessary small pressure gradient near the surface of the star will be reproduced if the value of the length parameter $a$ is small compared with the radius $R$ of the star. This length parameter also fixes the position of the minimum of $\mathrm{d}P/\mathrm{d}r$ at a distance $r = a/\sqrt{2}$ from the centre of the star.

The pressure inside the star is obtained by integrating Eq. (5.24) and imposing the boundary condition of zero pressure at $r = R$. This gives

$$P(r) = \frac{2\pi}{3} G\rho_c^2 a^2 [\exp(-r^2/a^2) - \exp(-R^2/a^2)]. \tag{5.25}$$

This representation for the pressure inside a star defines a family of stellar models, each specified by particular values of $\rho_c$, $a$ and $R$. The corresponding expressions for the density and temperature can be found as follows:

### Density and temperature inside a star

In order to find expressions for the density and temperature, we first calculate $m(r)$, the mass of stellar material enclosed by a sphere of radius $r$. To do so we combine Eqs. (5.1) and (5.2) to give

$$Gm(r)\,\mathrm{d}m = -4\pi r^4\,\mathrm{d}P,$$

which can be integrated to yield

$$G\frac{1}{2}m^2(r) = -4\pi \int_0^r r'^4 \frac{dP}{dr'}dr'.$$

If we substitute the expression (5.24) for the pressure gradient, we obtain

$$m(r) = \frac{4\pi a^3}{3}\rho_c \Phi(x), \tag{5.26}$$

where $x$ is $r/a$ and

$$\Phi^2(x) = 6\int_0^x x'^5 \exp(-x'^2)dx' = 6 - 3(x^4 + 2x^2 + 2)\exp(-x^2). \tag{5.27}$$

Given this expression for $m(r)$, it is easy to find the density and the temperature inside the star. The density $\rho(r)$ can be found directly; it is given by

$$\rho(r) = \frac{1}{4\pi r^2}\frac{dm}{dr} = \rho_c \left[\frac{x^3 \exp(-x^2)}{\Phi(x)}\right]. \tag{5.28}$$

The temperature $T(r)$ requires a knowledge of the equation of state of the stellar material. For example, if we assume that the star is supported by an ideal classical gas we can use Eqs. (5.8) and (5.9) to give

$$T(r) = \frac{\overline{m}}{k}\frac{P(r)}{\rho(r)} \quad \text{with} \quad \overline{m} = \frac{2m_{\rm H}}{[1 + 3X_1 + 0.5X_4]}. \tag{5.29}$$

The density and temperature distributions given by Eqs. (5.28) and (5.29) are expected to be more reliable at small $r$ where the prescription (5.24) accurately reproduces the pressure gradient. In particular, we can use the small $x$ behaviour of the function $\Phi(x)$,

$$\Phi(x) = \left[x^6 - \frac{3}{4}x^8 + \frac{3}{10}x^{10} - \frac{1}{12}x^{12} + ldots\right]^{1/2}, \tag{5.30}$$

to derive the following expressions for the density and temperature at small $r$:

$$\rho(r) = \rho_c\left[1 - \frac{5}{8}\frac{r^2}{a^2} + \ldots\right] \quad \text{and} \quad T(r) = T_c\left[1 - \frac{3}{8}\frac{r^2}{a^2} + \ldots\right]. \tag{5.31}$$

**A star with a high central density**

The model simplifies considerably if the mass of the star is concentrated towards the centre, so that the central density is much larger than the average density. If this is the case, the length parameter $a$ is small compared with the stellar radius $R$, and terms proportional to $\exp(-a^2/R^2)$ can be neglected. We note that this is a reasonable approximation when the model is applied to the sun; in the next section we shall show that in this case $a = R_\odot/5.4$.

In the simplified model with small $a$, the total mass of the star is simply

$$M = m(R) = \frac{4\pi\rho_c a^3}{3}\Phi(R/a) \approx \frac{4\pi\rho_c a^3\sqrt{6}}{3}. \tag{5.32}$$

It follows that the average density of the star is about $\sqrt{6}(a/R)^3\rho_c$. Furthermore, it is straightforward to show that the density at $r = a$ is $0.53\rho_c$, and that a sphere of radius $a$ contains 28% of the mass of the star. In addition, we can obtain a very useful relation between the pressure and the density at the centre of the star. Substituting $r = 0$ into Eq. (5.25) gives a central pressure of the form

$$P_c \approx \frac{2\pi}{3}G\rho_c^2 a^2.$$

If we use Eq. (5.32) to express $a$ in terms of $M$ and $\rho_c$, we find

$$P_c \approx \left[\frac{\pi}{36}\right]^{1/3} GM^{2/3}\rho_c^{4/3}, \tag{5.33}$$

where the numerical factor $[\pi/36]^{1/3}$ is approximately 0.44.

This equation predicts a relation between the pressure and density at the centre of a star which is expected to be approximately valid for any homogeneous star in which the mass is concentrated towards the centre. Moreover, the relation does not depend on the specific value of the parameter $a$, as long as it is small. We note here that other models for stellar structure give a similar relation. For example, Eq. (5.33) is roughly consistent with polytrope models of stellar structure. A polytrope model with index $n$ is defined by assuming Eq. (5.17) and solving the differential equation (5.18). It can be shown that a polytrope with index $n = 3/2$ gives

$$P_c = 0.48\ GM^{2/3}\rho_c^{4/3}, \tag{5.34}$$

and a polytrope with index $n = 3$ gives

$$P_c = 0.36\ GM^{2/3}\rho_c^{4/3}. \tag{5.35}$$

We also refer the reader to problem 1.7 at the end of Chapter 1. In the last part of this problem you are asked to show that under very general conditions there is an upper bound for the central pressure given by

$$P_c < \left[\frac{\pi}{6}\right]^{1/3} GM^{2/3}\rho_c^{4/3}. \tag{5.36}$$

Thus, in many situations, the central pressure needed to support a star is approximately given by Eq. (5.33). This equation provides a simple and moderately reliable way of imposing the condition of hydrostatic equilibrium in a stellar structure calculation. We shall use it in Section 5.4 to derive estimates for the minimum and maximum masses for stars. It will also be used in the analysis of white dwarfs in Section 6.1 of Chapter 6.

## 5.3 MODELLING THE SUN

Heat transfer and thermonuclear fusion are the essential ingredients of a model of the sun. Realistic solar models take careful account of the chemical composition, and the changes in the composition as the sun evolves. This chemical composition determines the opacity of the matter within the sun, with some elements of very low abundance having a large effect. Energy transport is usually governed by radiative diffusion, but convective transport dominates in a zone near to the solar surface. The equation of state, i.e. the relation between the pressure, temperature and density, takes into account the effects of electron degeneracy which begin to become significant near the solar centre; the equation of state also takes into account the pressure due to photons or radiation. Finally, thermonuclear energy is produced by the reactions of the proton–proton chain and, to a lesser extent, by the reactions of the carbon–nitrogen cycle.

The Standard Solar Model is widely recognized as one of the most realistic models of the sun. It provides a framework for the intrepretation of all observational properties, including the flux of solar neutrinos. A full description of this detailed and sophisticated model is given by Bahcall (1989).

But the purpose of this section is less ambitious. It is to obtain a rough understanding of the sun in terms of a simple model based upon the prescription (5.34) for the pressure gradient. This prescription is not appropriate for a model of today's sun. Hydrogen burning during the last 5 billion years has led to a composition which changes abruptly near the centre of the sun. As a result, there is an abrupt change in the density and in the pressure gradient. However, the prescription for the pressure gradient given by Eq. (5.34) is a reasonable starting point for a model of the early, chemically homogeneous sun. Such a simple model is very crude by the standards set by realistic models. But despite this, we shall see that it gives a useful insight into the variation of the pressure, density and temperature inside the sun. Moreover, we shall show that it yields the correct order of magnitude for the solar luminosity in two independent ways, firstly by assuming the release of nuclear energy by the reactions of the proton–proton chain, and secondly by assuming heat transport by radiative diffusion.

### Pressure, density and temperature

Following Clayton (1986), we shall compare the predictions given by the simple model with the numerical results of a stellar structure calculation by Strömgren;

these numerical results for a chemically homogeneous sun are tabulated in Table 6.5 of Clayton (1983). In particular, Strömgren obtained the following results for the pressure, density and temperature at the solar centre:

$$P_c = 1.65 \times 10^{16} \text{ Pa}, \quad \rho_c = 9.0 \times 10^4 \text{ kg m}^{-3} \quad \text{and} \quad T_c = 13.7 \times 10^6 \text{ K}, \tag{5.37}$$

As indicated by Eq. (5.25), the simple model is specified by three parameters: the central density $\rho_c$, the length parameter $a$ and the radius $R$. For a star with a high central density we can use Eq. (5.32) to express the length parameter $a$ in terms of the mass $M$ and central density $\rho_c$. Thus to model the sun we take $R = R_\odot$ and $M = M_\odot$, and $\rho_c$ equal to Strömgren's value of $9 \times 10^4$ kg m$^{-3}$. This implies that the length parameter is $a = R_\odot/5.4$.

According to Eq. (5.33), the central pressure needed to support a star of high central density $\rho_c$ and mass $M_\odot$ is

$$P_c = 0.44\, G M_\odot^{2/3} \rho_c^{4/3}. \tag{5.38}$$

This gives a central pressure $P_c = 1.9 \times 10^{16}$ Pa, which is slightly higher than Strömgren's value of $1.65 \times 10^{16}$ Pa. To calculate the central temperature we assume that the matter at the centre of the sun is an ideal classical gas with hydrogen and helium mass fractions of $X_1 = 0.71$ and $X_4 = 0.27$. This assumption yields a central temperature $T_c = 16 \times 10^6$ K.

To find the variation in the pressure, density and temperature inside the sun, we adopt the appropriate value for the length parameter, $a = R_\odot/5.4$, use Eqs. (5.25), (5.28) and (5.29) to give the results illustrated in Fig. 5.2. This figure shows impressive agreement between the results obtained from Clayton's simple model and those obtained by Strömgren by numerical solution of the equations of stellar structure.

### The solar luminosity

The temperature and density distributions illustrated in Fig. 5.2 can be used to estimate the luminosity of the sun in two independent ways. First, we can integrate Eq. (5.4) and relate the luminosity to the total power due to thermonuclear fusion. Second, we can use Eq. (5.3) and estimate the power flow that can be achieved by radiative diffusion. These two estimates should agree with each other. Indeed, in practice, the highly temperature-dependent nuclear reactions in the sun adjust themselves so that the nuclear power generated equals the power lost by radiative diffusion towards the surface. Furthermore, these estimates for the luminosity should be comparable to the computed luminosity, which, for the early homogeneous sun, is about $3 \times 10^{26}$ W.

To find the solar luminosity due to thermonuclear fusion we assume that nuclear energy is generated by the reactions of the proton–proton chain in accordance with Eq. (5.15). In fact, the power produced by the proton–proton chain was slightly

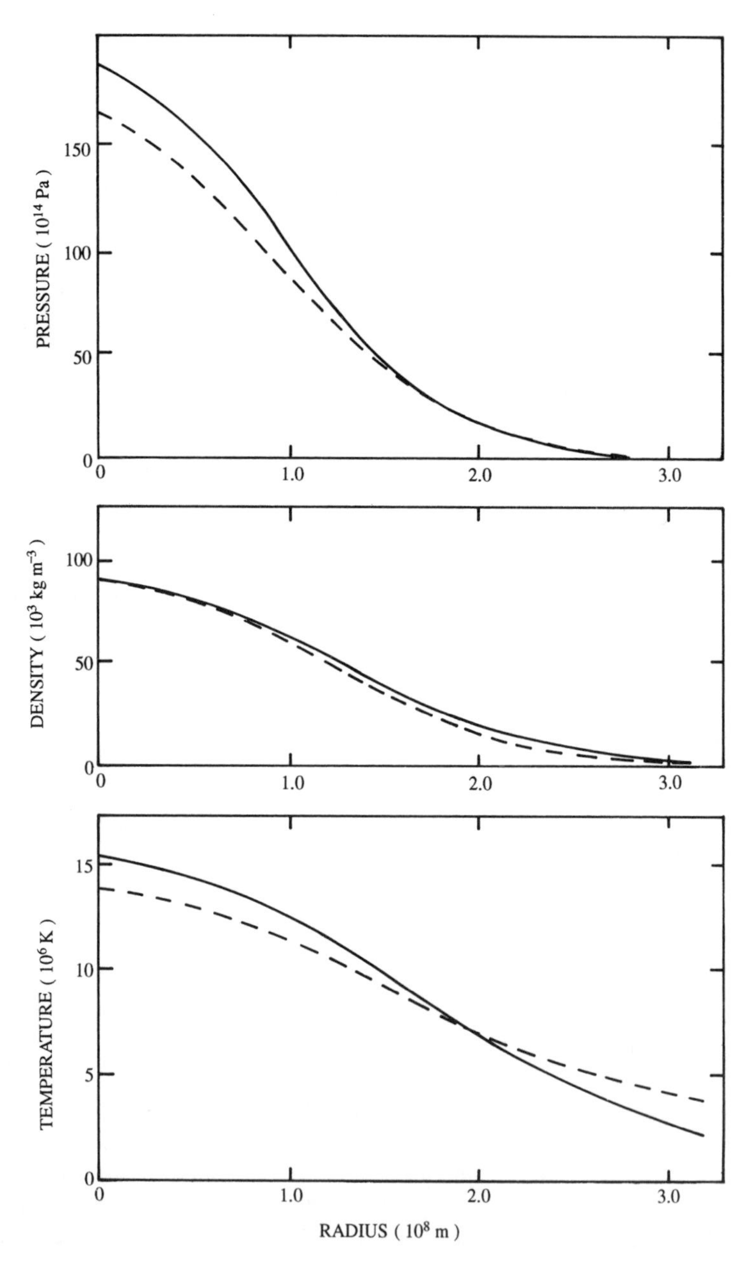


Fig. 5.2 The pressure, density and temperature in a homogeneous sun with $X_1 = 0.71$ and $X_4 = 0.27$. The broken lines represent the results of a computer solution by B. Strömgren; see Table 6.5 of Clayton (1983). The solid lines correspond to the simple model described in Section 5.2 with $a = R_{\odot}/5.4 = 1.29 \times 10^8$ m.

lower in the early sun because branch II of the chain was less effective when the helium-four abundance was lower. If we substitute a slightly modified Eq. (5.15) into Eq. (5.4) and integrate, we obtain

$$L_{\odot} = 8.4 \times 10^{-37} X_1^2 \int_0^{R_{\odot}} 4\pi r^2 [\rho(r)]^2 [T(r)]^4 \mathrm{d}r. \tag{5.39}$$

If the density and temperature are given by Eqs. (5.28) and (5.29), we find

$$L_{\odot} = 8.4 \times 10^{-37} X_1^2 \, 4\pi a^3 \rho_c^2 T_c^4 \, I, \tag{5.40}$$

where $I$ is the integral

$$I = \int_0^{R_{\odot}/a} \frac{\Phi^2(x) \exp(-2x^2)}{x^4} \mathrm{d}x.$$

To evaluate this integral we use the expansion (5.30) to give

$$I = \int_0^{R_{\odot}/a} \left[ x^2 - \frac{3}{4} x^4 + \frac{3}{10} x^6 - \cdots \right] \exp(-2x^2) \mathrm{d}x.$$

If $a$ is small compared with $R_{\odot}$, the upper limit of the integral may be extended to infinity to give a series of integrals, which can be evaluated by using the standard integral

$$\int_0^{\infty} \exp(-\alpha x^2) \mathrm{d}x = (\pi/4\alpha)^{1/2},$$

and the integrals obtained by differentiation with respect to $\alpha$. If the series is summed numerically, we find that

$$I = 0.078 \, (\pi/2)^{1/2}. \tag{5.41}$$

Substitution of this result into Eq. (5.40) gives

$$L_{\odot} = 1.0 \times 10^{-36} a^3 X_1^2 \rho_c^2 T_c^4.$$

If we use Eq. (5.32) to express $a$ in terms of the mass and central density of the sun, we find

$$L_{\odot} = 1.0 \times 10^{-36} \frac{3}{4\pi\sqrt{6}} M_{\odot} X_1^2 \rho_c T_c^4. \tag{5.42}$$

In order to obtain a numerical value, we insert $X_1 = 0.71$, $\rho_c = 9.0 \times 10^4$ kg m$^{-3}$ and $T_c = 16 \times 10^6$ K into Eq. (5.42) to give a solar luminosity of about $5 \times 10^{26}$ W. This estimate based on thermonuclear fusion by the proton–proton chain should be compared with $3 \times 10^{26}$ W, the luminosity for the early, homogeneous sun given by computer calculations.

Heat transport by radiative diffusion provides a second way of estimating the solar luminosity. Only a rough estimate can be obtained because the model makes no attempt to constrain the temperature gradient to ensure that the heat flow is in accordance with the power generation within in the sun; for example, the divergence of the heat flow should go to zero as the temperature and power generation fall off with increasing distance from the centre. Moreover, the model becomes increasingly poor at large distances and the power flow $L(r)$ can only be equated to $L_\odot$ for values of $r$ beyond the central power generating region. Nevertheless, we shall see that the model can still give a useful estimate for the solar luminosity due to the radiative diffusion of heat.

The relation between the temperature gradient and the power flow due to radiative diffusion in a star is given by Eq. (5.3). In order to avoid confusion between the length parameter $a$ of the simple model and the radiation constant, which is also denoted by $a$, we express the latter in terms of Stefan's constant $\sigma$ which equals $ac/4$. Rearranging Eq. (5.3) then gives an outward power flow at radius $r$ of the form

$$L(r) = -\frac{16\sigma}{3}\frac{4\pi r^2[T(r)]^3}{\kappa(r)\rho(r)}\frac{\mathrm{d}T}{\mathrm{d}r}. \tag{5.43}$$

We shall consider power flow in a central region of the sun where $r$ is small so that the density and temperature are given by Eq. (5.31). If the opacity obeys Kramers' law, Eq. (5.14), its value at a small distance $r$ from the centre of the sun is given by

$$\kappa(r) \approx \kappa_c \left[1 + \frac{11}{16}\frac{r^2}{a^2}\right],$$

where $\kappa_c$ is the opacity at the centre of the sun. The temperature gradient can be obtained by differentiating the power expansion for $T$ given in Eq. (5.31). If we only retain terms of order $r^2/a^2$ we find that Eq. (5.43) yields the following approximation for the power flow at a small distance $r$ from the centre of the sun:

$$L(r) \approx 16\pi\sigma\frac{T_c^4}{\kappa_c\rho_c}\frac{r^3}{a^2}\left[1 - \frac{19}{16}\frac{r^2}{a^2}\right]. \tag{5.44}$$

If we insert $\rho_c = 9.0 \times 10^4$ kg m$^{-3}$, $T_c = 16 \times 10^6$ K and the appropriate value for the opacity at this density and temperature, $\kappa_c = 0.14$ m$^2$ kg$^{-1}$, and use $a = R_\odot/5.4$ we obtain

$$L(r) \approx 3 \times 10^{29} \frac{r^3}{R_\odot^3} \left[ 1 - 35 \frac{r^2}{R_\odot^2} \right] \text{W}. \tag{5.45}$$

As expected, the power flow initially increases with $r$, but it does not approach a constant value as it would in a more realistic model. Despite this shortcoming, it is encouraging to note that, when $r$ is $R_\odot/10$, the outward power flow given by Eq. (5.45) reaches $2 \times 10^{26}$ W. Thus, the order of magnitude of the power flow due to radiative diffusion is comparable with $5 \times 10^{26}$ W, our estimate of the solar luminosity due to thermonuclear fusion. In a realistic model of the sun, the power generated by thermonuclear fusion is precisely that necessary to supply the power transported towards the surface. If it were not, the sun would contract or expand until it is.

## 5.4 MINIMUM AND MAXIMUM MASSES FOR STARS

In practice most main sequence stars have a mass in the range from about a tenth of a solar mass to about fifty solar masses. Two questions immediately arise: what fundamental constants of nature determine the order of magnitude of the mass of a main sequence star? And why is the range in mass so limited? These questions were briefly considered in Chapter 1 and will now be considered in more detail. In this section, we shall see that if the mass is significantly smaller than the solar mass, gravitational contraction will result in an internal temperature which is insufficient to ignite thermonuclear fusion and create a genuine star. We shall also see that if the mass of the star greatly exceeds the solar mass, then radiation pressure becomes dominant. As a result, the binding energy of the star is small and any small energy loss or gain is accompanied by large changes in the thermal kinetic energy and gravitational potential energy; in other words, the hydrostatic equilibrium of the star becomes precarious.

The key ingredient in the calculation of minimum and maximum stellar masses is the condition for hydrostatic equilibrium. We shall impose this condition in a simple and approximate way by focusing on the pressure at the centre of a star. To do so we recall Eq. (5.33). This states that the central pressure and density of a star of mass $M$ in hydrostatic equilibrium are related by

$$P_c \approx \left[ \frac{\pi}{36} \right]^{1/3} G M^{2/3} \rho_c^{4/3}, \tag{5.46}$$

a relation which is approximately true for any chemically homogeneous star, in which the mass is concentrated towards the centre so that the central density greatly exceeds the average density. The pressure given by Eq. (5.46) is the central pressure needed to support the star. We shall see that expressions for the minimum and maximum masses of stars can be found by examining the source of this pressure.

**Minimum mass of a main sequence star**

In order to achieve stardom a contracting system must be sufficiently massive to generate a central temperature which is high enough for thermonuclear fusion to supply the energy loss from the surface. To derive the minimum mass needed to reach this ignition temperature, we consider a contracting cloud of ionized gas with mass $M$.

Initially, the energy lost from the surface is supplied by gravitational contraction. The pressure is low and, to a first approximation, the electrons and ions form an ideal classical gas so that the central pressure and temperature are related by

$$P_c = \frac{\rho_c}{\overline{m}} kT_c, \tag{5.47}$$

where $\overline{m} = 2m_{\rm H}/[1 + 3X_1 + 0.5X_4]$ is the average mass of the gas particles; see Eqs. (5.8) and (5.9). If the pressure generated by the ideal gas is close to the pressure needed to support the system, the contraction is slow and the cloud is in a state close to hydrostatic equilibrium. Equating the pressures given by Eqs. (5.46) and (5.47) gives the following expression for the central temperature during this period of slow contraction:

$$kT_c \approx \left[\frac{\pi}{36}\right]^{1/3} G\overline{m}M^{2/3}\rho_c^{1/3}. \tag{5.48}$$

We see that the temperature rises steadily as the density of the contracting gas cloud increases.

The temperature of a contracting cloud will continue to rise until either a substantial amount of energy is released by thermonuclear fusion, or the electrons at the centre become degenerate. In the former case, nuclear energy alone can supply the energy loss from the surface, thereby removing the need for contraction and the release of gravitational energy. In the latter case, electrons, occupying the lowest possible energy states in accordance with the Pauli exclusion principle, resist compression and support the mass. Thus, true stardom will not be possible if the electrons become degenerate before the ignition temperature for thermonuclear fusion is reached.

To estimate the maximum temperature achievable at the centre of a contracting gas cloud, we shall assume a stage is reached in which the electrons at the centre are fully degenerate and the ions are classical. At this stage the central pressure is given by

$$P_c = K_{NR} n_e^{5/3} + n_i kT_c, \tag{5.49}$$

where the constant $K_{NR}$ is given by Eq. (5.10). The number densities for the electrons and ions can be expressed in terms of the central density using Eqs. (5.6) and (5.7). However in order to simplify the algebra, we shall assume the mass is entirely composed of hydrogen so that $n_e = n_i = \rho_c/m_{\rm H}$. In this case

$$P_c = K_{NR}\left[\frac{\rho_c}{m_{\rm H}}\right]^{5/3} + \frac{\rho_c}{m_{\rm H}}kT_c. \tag{5.50}$$

Again, hydrostatic equilibrium is achieved if this pressure equals the pressure needed to support the mass. Equating the pressure given by Eqs. (5.50) to the pressure given by Eq. (5.46) leads to a central temperature given by

$$kT_c \approx \left[\frac{\pi}{36}\right]^{1/3} Gm_{\rm H}M^{2/3}\rho_c^{1/3} - K_{NR}\left[\frac{\rho_c}{m_{\rm H}}\right]^{2/3}. \tag{5.51}$$

Equation (5.51) gives the central temperature in a contracting mass of hydrogen at a stage when the electrons at the centre are fully degenerate and the ions are classical. In contrast with Eq. (5.48), there are two terms. The first is associated with the classical ions and the second with the degenerate electrons. The second term becomes important at high density and when it does the temperature will cease to rise quickly as the mass contracts. This behaviour is illustrated in Fig. 5.3 which shows how the temperature at the centre of a contracting cloud with mass $M_\odot/16$ varies as the density increases. As expected, the temperature initially rises as $\rho_c$ increases. However, the temperature increases less quickly as the pressure due to degenerate electrons becomes more important; eventually the degeneracy pressure is dominant and the temperature ceases to rise. To find the maximum value of the temperature, we rewrite Eq. (5.51) in the form

$$kT_c = A\rho_c^{1/3} - B\rho_c^{2/3}.$$

Elementary calculus then shows that $kT_c$ reaches a maximum of $A^2/4B$ at a density of $(A/2B)^3$. Substituting for $A$ and $B$, we find that the maximum temperature reached at the centre of a contracting mass of hydrogen is

$$[kT_c]_{max} \approx \left[\frac{\pi}{36}\right]^{2/3} \frac{G^2 m_{\rm H}^{8/3}}{4K_{NR}} M^{4/3}. \tag{5.52}$$

A less accurate version of this equation was derived in Chapter 1; see Eq. (1.28).

We can now impose the condition that the contracting mass achieves stardom. This condition is that the maximum central temperature reaches the ignition temperature for the thermonuclear fusion of hydrogen. If we denote this ignition temperature by $T_{ign}$, we find that the minimum mass for a genuine star is given by

$$M_{min} \approx \left[\frac{36}{\pi}\right]^{1/2} \left[\frac{4K_{NR}}{G^2 m_{\rm H}^{8/3}}\right]^{3/4} \left[kT_{ign}\right]^{3/4}. \tag{5.53}$$

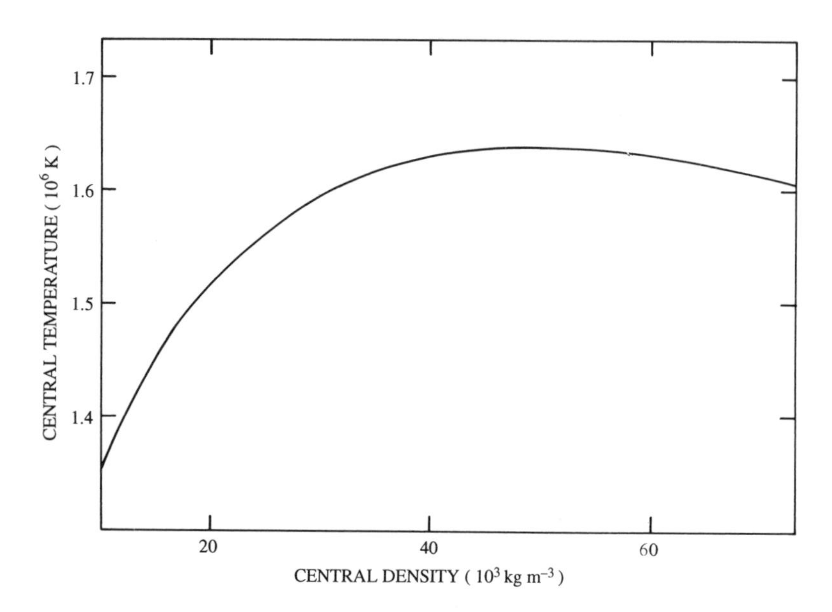


Fig. 5.3 The temperature at the centre of a contracting cloud of hydrogen with mass $M_\odot/16$ as a function of the central density.

In fact, the ignition temperature of a material depends on its environment. It is the temperature at which the power produced in a particular region begins to match the power that escapes from the region. When this occurs the region gets hot and the 'fire' spreads. If we take the ignition temperature for hydrogen to be about $1.5 \times 10^6$ K, one-tenth of the central temperature of the sun, Eq. (5.53) gives a value of $0.05\ M_\odot$ for the minimum mass of a star; more accurate calculations give values closer to $0.1\ M_\odot$.

## Maximum mass of a main sequence star

We recall from Section 1.2 of Chapter 1 that the hydrostatic equilibrium of a star becomes precarious if the pressure preventing gravitational contraction is supplied by a gas of ultra-relativistic particles. This implies that a star could be easily disrupted if radiation becomes the dominant source for the internal pressure. This general property sets an upper limit to the mass of a main sequence star. To derive this limit we consider the pressure due to electrons, ions and photons at the centre of a hot, massive star and compare this with the pressure needed to support the star.

We assume that the electrons, ions and photons are in thermal equilibrium at a temperature $T_c$ and a density $\rho_c$ at the centre of the star. The central pressure $P_c$ is the sum of a gas pressure $P_g$ due to electrons and ions, and a radiation pressure $P_r$ due to photons.[1] At high temperature and low density, the electrons and ions form an ideal classical gas with a gas pressure given by

$$P_g = \frac{\rho_c}{\overline{m}} k T_c. \tag{5.54}$$

According to Eq. (5.12), the photons form an ideal quantum gas with pressure

$$P_r = \frac{1}{3} a T_c^4. \tag{5.55}$$

It is convenient to describe the fractional contributions of gas and radiation pressure to the total pressure $P_c = P_r + P_g$ by introducing the parameter $\beta$. We take

$$P_g = \beta P_c \quad \text{and} \quad P_r = (1 - \beta) P_c. \tag{5.56}$$

It is then straightforward to eliminate $T_c$ from Eqs. (5.55) and (5.54) and express $P_c$ in terms of $\beta$ and $\rho_c$ to give

$$P_c = \left[ \frac{3}{a} \frac{(1 - \beta)}{\beta^4} \right]^{1/3} \left[ \frac{k \rho_c}{\overline{m}} \right]^{4/3}. \tag{5.57}$$

Again, hydrostatic equilibrium is achieved if this pressure equals the pressure needed to support the star. Equating the pressure given by Eq. (5.57) to the pressure given by Eq. (5.46) leads to

$$\left[ \frac{\pi}{36} \right]^{1/3} G M^{2/3} = \left[ \frac{3}{a} \frac{(1 - \beta)}{\beta^4} \right]^{1/3} \left[ \frac{k}{\overline{m}} \right]^{4/3} \tag{5.58}$$

We recall that $(1 - \beta)$ and $\beta$ are the fractional contributions of 'radiation' and 'gas' to the central pressure; by definition both are less than one. We note from Eq. (5.58) that $M$, the mass of the star, determines $\beta$ and that $\beta$ decreases as $M$ increases. Hence the radiation pressure $P_r = (1 - \beta) P_c$ is more important in stars with a large mass. We illustrate this in Fig. 5.4, which plots $P_r / P_c$ as a function of the mass of the star; a value of 0.61 amu has been taken for $\overline{m}$.

We recall from Chapter 1 that the binding energy is small for a star supported by a pressure due to the random motion of ultra-relativistic particles such as photons. Moreover, the release or the absorption of a small amount of energy in such a star is accompanied by large changes in the internal kinetic energy and gravitational potential energy. It follows that radiation pressure has a destabilizing effect on

[1] Of course radiation pressure can be thought of as a pressure due to a gas of photons. Despite this we will follow custom at this stage and use the adjective *gas* to describe the pressure due to electrons and ions.

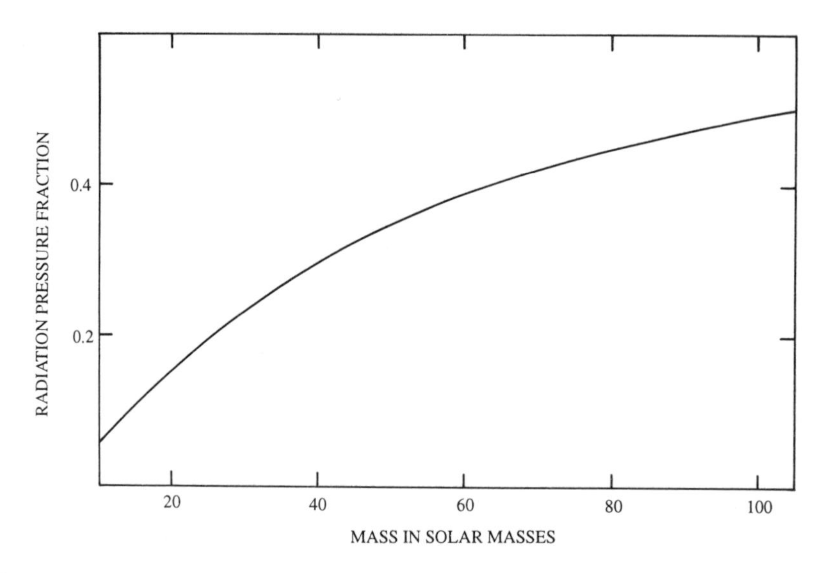


Fig. 5.4 The fractional contribution of radiation pressure to the gas pressure at the centre of a star of mass $M$. Note that radiation pressure becomes increasingly important in more massive stars.

massive stars. In practice, the increasing importance of radiation pressure in massive stars illustrated in Fig. 5.4 imposes an upper limit on the mass of main sequence stars. An estimate of about $100 M_{\odot}$ for the maximum mass of a main sequence star can be obtained from Eq. (5.58) by requiring $(1 - \beta)$ to be less than 0.5; this corresponds to assuming that not more than 50% of the pressure at the centre of the star is due to radiation. In fact, stars with a mass greater than $50\ M_{\odot}$ are very rare.

## A fundamental unit for stellar masses

We have seen how lower and upper limits for stellar masses are imposed by the need for thermonuclear fusion and the destabilizing effect of radiation pressure. A lower limit can be found from Eq. (5.53) and an upper limit from Eq. (5.58). The range of masses for main sequence stars is surprisingly small, typically from about one-tenth of a solar mass to about fifty solar masses. Thus, the solar mass seems to be a convenient unit for the mass of all main sequence stars. We shall now re-examine these results and identify the fundamental constants of nature which actually fix the mass of a main sequence star in the region of a solar mass.

To begin with, we shall introduce a dimensionless measure of the strength of the gravitational interaction between two nucleons. Because the average mass of a neutron and proton is almost equal to the mass of a hydrogen atom, we shall denote the nucleon mass by $m_{\mathrm{H}}$. The gravitational potential energy of two nucleons at a distance $r$ is $-Gm_{\mathrm{H}}^2/r$. The magnitude of this energy for nucleons separated by a fundamental distance, $\hbar/m_{\mathrm{H}}c$, in units of the fundamental rest-mass energy, $m_{\mathrm{H}}c^2$, is

$$\alpha_G = \frac{Gm_{\mathrm{H}}^2}{\hbar c} = 5.9 \times 10^{-39}. \tag{5.59}$$

This small dimensionless number is a measure of the strength of the gravitational interaction between nucleons. A corresponding measure for the strength of the electromagnetic interaction is the fine structure constant, $\alpha = e^2/(4\pi\epsilon_0\hbar c) = 1/137$. Because we have usually used Planck's constant in this book, we point out that the definitions of $\alpha_G$ and $\alpha$ involve $\hbar$ or $h/2\pi$.

The minimum mass of a main sequence star is given by Eq. (5.53). This equation contains the constant $K_{NR}$ which depends on Planck's constant and the mass of the electron as shown in Eq. (5.10). If we use Eq. (5.10), we find that the minimum stellar mass can be rewritten as

$$M_{min} \approx 16 \left[\frac{kT_{ign}}{m_e c^2}\right]^{3/4} \alpha_G^{-3/2} m_{\mathrm{H}}. \tag{5.60}$$

If we take $T_{ign}$ to be about $1.5 \times 10^6$ K, one-tenth of the central temperature of the sun, we find

$$M_{min} \approx 0.03\ \alpha_G^{-3/2} m_{\mathrm{H}}. \tag{5.61}$$

An estimate for the maximum mass of a main sequence star can be obtained from Eq. (5.58) by requiring $(1-\beta)$ to be 0.5; this corresponds to assuming that 50% of the pressure at the centre of the star is due to radiation. Using Eq. (5.12) to relate the radiation constant $a$ to Planck's constant and the velocity of light, and assuming that the average mass $\overline{m}$ is 0.61 $m_{\mathrm{H}}$, we find a maximum mass given by

$$M_{max} \approx 56\ \alpha_G^{-3/2} m_{\mathrm{H}}. \tag{5.62}$$

In view of Eqs. (5.61) and (5.62), we introduce the mass

$$M_\star = \alpha_G^{-3/2} m_{\mathrm{H}} = 1.85\ M_\odot, \tag{5.63}$$

and identify this mass as a fundamental stellar mass which determines the mass scale of main sequence stars. It depends solely on the mass of the nucleon and the dimensionless strength of the gravitational interaction between nucleons. A stable, long-lived main sequence star occurs if $M \approx M_*$, thermonuclear fusion is not ignited if $M << M_\star$, and radiation pressure destabilizes if $M >> M_\star$. In view of this, it is no accident that the mass of the sun is comparable with $M_\star$. Finally, the role of $M_\star$ as a fundamental stellar mass indicates that the number of nucleons in a typical star is solely determined by $\alpha_G$. This number is

$$N_\star = \frac{M_\star}{m_\mathrm{H}} = \alpha_G^{-3/2} = 2 \times 10^{57}. \tag{5.64}$$

## SUMMARY

### Preamble

- Stellar structure calculations are based on four fundamental equations, Eqs. (5.1) to (5.4), which describe hydrostatic equilibrium, mass conservation, heat transport and power generation within a star. These equations are differential equations for four unknown functions $P(r)$, $m(r)$, $T(r)$ and $L(r)$. They can be solved, in principle, if they are supplemented by an equation of state for the stellar material and expressions for the opacity and power generation.

### A simple model for a star

- Insight into some of the features of stellar structure can be obtained by assuming the following simple analytic form for the pressure gradient:

$$\frac{\mathrm{d}P}{\mathrm{d}r} = -\frac{4\pi}{3} G\rho_c^2 r \exp(-r^2/a^2). \tag{5.24}$$

This expression describes the pressure gradient correctly at small $r$ and very approximately at large $r$. Once the mass, radius and central density of the star are specified, the parameter $a$ is fixed and simple expressions can then be obtained for the density within the star; see Eq. (5.28). The temperature distribution can also be found if the equation of state is known; see Eq. (5.29).
- Whenever the mass of the star tends to concentrate towards the centre, there is a simple relation between the pressure and the density at the centre of the star given by

$$P_c \approx \left[\frac{\pi}{36}\right]^{1/3} G M^{2/3} \rho_c^{4/3}. \tag{5.33}$$

Other models give very similar relations. This pressure–density relation can be used as a simple and direct way of imposing the condition of hydrostatic equilibrium, as in Section 5.4, in order to estimate the minimum and maximum masses of main sequence stars.

### Modelling the sun

- The pressure, density and temperature within the sun can be roughly reproduced by assuming a structure derived from the pressure gradient (5.24); see Fig. 5.2.
- The solar luminosity can be estimated by considering thermonuclear fusion, Eq. (5.42), and by considering heat transport by radiative diffusion, Eq. (5.45). In both cases the estimates are comparable with the observed luminosity.

### Minimum and maximum masses for stars

- A fundamental stellar mass can be defined by

$$M_\star = \alpha_G^{-3/2}\, m_{\rm H} = 1.85\ M_\odot, \tag{5.63}$$

  where $\alpha_G$ is a dimensionless measure of the strength of the gravitational interaction between two nucleons, $\alpha_G = Gm_{\rm H}^2/\hbar c$. This fundamental stellar mass is the natural unit for all stellar masses. It corresponds to a star containing $2 \times 10^{57}$ nucleons; see Eq. (5.64).
- A contracting cloud of hydrogen achieves true stardom and ignites the thermonuclear fusion of hydrogen if its mass is greater than a mass given by Eq. (5.53). This minimum mass corresponds to

$$M_{min} \approx 0.03\ M_\star. \tag{5.61}$$

- If the mass of a star exceeds a maximum value given by

$$M_{max} \approx 56\ M_\star, \tag{5.62}$$

  the internal radiation pressure dominates the gas pressure, and the hydrostatic equilibrium of the star becomes precarious.

## PROBLEMS 5

5.1 Consider a star of mass $M$ and radius $R$ in which the pressure gradient is given by

$$\frac{\mathrm{d}P}{\mathrm{d}r} = -\frac{4\pi}{3} G\rho_c^2 r \exp(-r^2/a^2),$$

where $a$ is a length parameter and $\rho_c$ is the central density; see Eq. (5.24). Derive an expression for the gravitational potential energy $E_{GR}$ of the star by using the Virial Theorem, Eq. (1.7). Show that if the length parameter $a$ is small compared with the radius $R$, the gravitational potential energy is approximately

$$E_{GR} \approx \frac{1}{3}\frac{R}{a}\frac{GM^2}{R}.$$

5.2 Consider a family of chemically homogeneous stars which are similar in every respect except for their masses and radii. The similarity of the stars is such that, for any member of the family with mass $M$ and radius $R$, the density at distance $r$ from centre can be written as a function of $x=r/R$ in the following way:

$$\rho(r) = \frac{M}{R^3} F_\rho(x),$$

where the function $F_\rho(x)$ is common to the entire family. In a similar way, the mass enclosed by a sphere of radius $r$ within the star can be written as

$$m(r) = M\, F_m(x),$$

where, again, the function $F_m(x)$ is common to the family.

Assume that the equation of state for the stellar material is the ideal classical gas equation, that the opacity of the material obeys Kramers' law, Eq. (5.14), and that nuclear energy is generated by the proton–proton chain in accordance with Eq. (5.15). Use the fundamental equations of stellar structure, (5.1) to (5.4), to derive the following scaling relations for the pressure, the temperature, the power flow due to radiative diffusion and the power flow due to nuclear fusion:

$$P(r) = \frac{M^2}{R^4} F_P(x),$$

$$T(r) = \frac{M}{R} F_T(x),$$

$$L_{rad}(r) = \frac{M^{5.5}}{R^{0.5}} F_{rad}(x),$$

$$L_{fus}(r) = \frac{M^6}{R^7} F_{fus}(x),$$

where, again, the functions are common to the family.

Note that the power flow due to radiative diffusion increases slowly and the power flow due to nuclear fusion increases rapidly as the star contracts. In fact, this rapid increase in the fusion power only occurs after the central temperature reaches a value in the neighbourhood of 10 million degrees. By sketching these power flows as a function of the radius $R$, illustrate how stars belonging to this family will contract until they reach radii and luminosities which are approximately given by

$$R \propto M^{0.077} \quad \text{and} \quad L \propto M^{5.45}.$$

Finally, show that this family of stars will lie on a line on the Hertzsprung–Russell diagram given by

$$L \propto T_E^{4.12},$$

where $T_E$ is the effective surface temperature.

5.3 Consider a family of stars in which the opacity is dominated by Thomson scattering by electrons, and in which nuclear energy is generated by the carbon–nitrogen cycle. This implies that the opacity is independent of the density and temperature (see Eq. (5.13)), and that the rate of nuclear energy production is proportional to $\rho^2\, T^{18}$ (see Section

4.2). In analogy with problem **5.2**, find, for this family of stars, a relation between the radius and the mass, and a relation between the luminosity and the mass. Find also the line on the Hertzsprung–Russell diagram describing the luminosity and effective surface temperature for these stars.

5.4 Under very general conditions the central pressure $P_c$ supporting a star of mass $M$ satisfies the inequality

$$P_c < \left[\frac{\pi}{6}\right]^{1/3} GM^{2/3}\rho_c^{4/3},$$

where $\rho_c$ is the central density; see problem **1.7** at the end of Chapter 1. Assume that part of this pressure, denoted by $\beta\, P_c$, is due to an ideal, classical gas of electrons and ions with average mass $\overline{m}$, and that the remaining pressure, denoted by $(1-\beta)\, P_c$, is due to radiation. Show that the above inequality can be used to derive an upper bound for the quantity $(1-\beta)\, \beta^4$. Use this bound to set limits on the fraction of the pressure due to radiation at the centre of stars of masses 1, 4 and 40 $M_\odot$.

# CHAPTER 6

# The end-points of stellar evolution

A star passes through several stages of nuclear burning each of which postpones gravitational contraction. It also loses weight by a variety of mechanisms by ejecting matter into outer space. Eventually, nuclear fusion at the centre of the star can no longer supply enough energy to sustain a high thermal pressure and the star contracts under gravity. A compact object is formed which can be a white dwarf, a neutron star or a black hole. We shall begin this chapter by considering white dwarfs, compact stars largely supported by the pressure of degenerate electrons. Most importantly, we shall show that the mass of a white dwarf cannot exceed the Chandrasekhar limit of about 1.4 $M_{\odot}$. We shall give an introduction to the physics of the neutron stars, compact objects which are largely supported by the pressure of degenerate neutrons, and consider the processes that lead to their formation. In particular, we shall address the crucial issue of the maximum mass of a neutron star and why we believe that all compact objects which exceed this mass must be completely collapsed objects whose only manifestation are intense gravitational fields. These disembodied remnants of matter are called black holes.

## 6.1 WHITE DWARFS

The sun will pass through its hydrogen burning phase and then a helium burning phase to form a star with a carbon–oxygen core surrounded by an envelope of helium and hydrogen. The temperature of the carbon–oxygen core will then increase as the core contracts under gravity. The increasing temperatures will accelerate the rate of helium burning in a shell surrounding the core and the envelope will expand and drift away to form a planetary nebula. But the contraction of the core is unlikely to result in the high temperature needed to burn carbon. The core, having lost its envelope, is expected to emerge as a hot white dwarf. As this white dwarf cools, the pressure generated by the thermal motion of the ions will become less important and eventually a pressure due to degenerate electrons will provide the bulk of the pressure needed to support the star.

### Mass and central density

We begin by considering the relation between the density at the centre of a white dwarf and its mass. To obtain this relation, we first write the number density for electrons at the centre of the star in terms of the central density $\rho_c$ in the following way:

$$n_e = Y_e \frac{\rho_c}{m_{\rm H}}, \tag{6.1}$$

where $Y_e$ is the number of electrons per nucleon; according to Eq. (5.6), $Y_e$ is approximately $[1+X_1]/2$. We now assume that the star is supported by the pressure of a gas of non-relativistic, degenerate electrons. This pressure is given by Eq. (5.10) which may be rewritten as

$$P = K_{NR} n_e^{5/3} = K_{NR} \left[ \frac{Y_e \rho_c}{m_{\rm H}} \right]^{5/3}. \tag{6.2}$$

If we equate this to the central pressure needed to support the star, which is given approximately by Eq. (5.33), we obtain

$$K_{NR} \left[ \frac{Y_e \rho_c}{m_{\rm H}} \right]^{5/3} \approx \left[ \frac{\pi}{36} \right]^{1/3} G M^{2/3} \rho_c^{4/3}. \tag{6.3}$$

Rearranging this equation and using the expression for $K_{NR}$ given in Eq. (5.10) leads to the following prediction for the density at the centre of a cold white dwarf of mass $M$:

$$\rho_c \approx \frac{3.1}{Y_e^5} \left[ \frac{M}{M_\star} \right]^2 \frac{m_{\rm H}}{(h/m_e c)^3}, \tag{6.4}$$

where $M_\star$ is the fundamental stellar mass defined by Eq. (5.63), i.e.

$$M_\star = \alpha_G^{-3/2} m_{\rm H} = 1.85\, M_\odot. \tag{6.5}$$

In deriving Eq. (6.4), we have assumed that the white dwarf is supported by the pressure of a gas of non-relativistic, degenerate electrons. However, we saw in Section 2.2 of Chapter 2 that degenerate electrons become relativistic when the number density of electrons is large compared with $(m_e c/h)^3$. In fact, the Fermi momentum of the electrons, which is given by Eq. (2.27), equals $m_e c$ when the number density is $(8\pi/3)(m_e c/h)^3$. We conclude that the electrons in the white dwarf will be relativistic if the density is large compared with $m_{\mathrm{H}}/(h/m_e c)^3$. It follows that the non-relativistic Eq. (6.4) can only be valid if the mass of the white dwarf, $M$, is small compared with $M_\star$.

As an example, we consider a carbon white dwarf of mass $0.4\ M_\odot$. The central density predicted by Eq. (6.4) is then $4.6\, m_{\mathrm{H}}/(h/m_e c)^3$, or about $5.4 \times 10^8$ kg m$^{-3}$. At this density, electrons have a Fermi momentum of $0.65\ m_e c$ and a Fermi kinetic energy of $0.19\ m_e c^2$. Hence the use of non-relativistic kinematics is, at best, a rough approximation. It is clear that for white dwarfs more massive than $0.4\ M_\odot$ one must completely take into account the effects of relativity in evaluating the pressure of the degenerate electron gas.

When relativity is taken into account, the calculated central density of a white dwarf is higher than that predicted by Eq. (6.4). In particular, the density, considered as a function of the white dwarf mass $M$, increases more rapidly than $M^2$. This arises because, as the density increases and the electrons become more relativistic, the equation of state is modified. In fact, when the electrons become ultra-relativistic at densities very large compared with $m_{\mathrm{H}}/(h/m_e c)^3$, the non-relativistic equation of state, Eq. (6.2), is replaced by

$$P = K_{UR} n_e^{4/3} = K_{UR} \left[ \frac{Y_e \rho_c}{m_{\mathrm{H}}} \right]^{4/3}, \tag{6.6}$$

where the constant $K_{UR}$ is determined by the fundamental constants $h$ and $c$ as indicated in Eq. (5.11). If this pressure due to ultra-relativistic electrons supports a star of mass $M$, then

$$K_{UR} \left[ \frac{Y_e \rho_c}{m_{\mathrm{H}}} \right]^{4/3} \approx \left[ \frac{\pi}{36} \right]^{1/3} G M^{2/3} \rho_c^{4/3}. \tag{6.7}$$

In the context of a model where a white dwarf is solely dependent on degenerate electrons for its support, this equation should be viewed as an equation for the mass of the white dwarf whose central density is very large compared with $m_{\mathrm{H}}/(h/m_e c)^3$, in effect a central density which tends to infinity. This mass is called the Chandrasekhar mass. Because the density cancels in Eq. (6.7), this mass is determined by $Y_e$ and fundamental constants like $G$, $m_{\mathrm{H}}$ and $K_{UR}$. It is given by

$$M_{CH} \approx \left[ \frac{36}{\pi} \right]^{1/2} \left[ \frac{Y_e}{m_{\mathrm{H}}} \right]^2 \left[ \frac{K_{UR}}{G} \right]^{3/2}. \tag{6.8}$$

If we express the constant $K_{UR}$ in terms of $h$ and $c$ by using Eq. (5.11) and recall the definition (6.5) of the fundamental stellar mass $M_{\star}$, we find that

$$M_{CH} \approx 2.3\, Y_e^2\, M_{\star} = 4.3\, Y_e^2\, M_{\odot}. \tag{6.9}$$

To understand the significance of the Chandrasekhar mass, consider a sequence of white dwarfs with increasing mass. As the mass increases, the degenerate electrons at the centre of the star become increasingly relativistic. When the mass is small, the central density will increase with mass in accordance with Eq. (6.4). As the mass becomes larger, the density increases more rapidly and, when the mass reaches $M_{CH}$, the density must approach infinity. In reality, the density becomes large compared with $m_{\rm H}/(h/m_e c)^3$, the star collapses and new physics must be sought to explain what happens next. For the moment, the only firm conclusion we draw is that a degenerate electron gas cannot support a star with mass larger than the Chandrasekhar mass.

The physical significance of the Chandrasekhar mass can be made clearer by considering a more general model for a white dwarf. So far we have considered two extreme models based on Eqs. (6.2) and (6.6); namely, a star supported by a low density gas of non-relativistic, degenerate electrons, and a star supported by a high density gas of ultra-relativistic, degenerate electrons. We shall now consider a model which incorporates both the non-relativistic and the ultra-relativistic extreme. To do so we consider a star supported by a degenerate gas of electrons whose kinematics are described by the exact relation between the energy and momentum,

$$\epsilon_p^2 = m_e^2 c^4 + p^2 c^2. \tag{6.10}$$

The pressure in such a gas can be obtained by using Eq. (2.13) and by noting that the velocity of a particle with momentum $p$ is $v_p = pc^2/\epsilon_p$. Because all the electrons in a degenerate gas fully occupy all the states with a momentum less than the Fermi momentum $p_F$, the pressure is given by

$$P = \frac{1}{3V}\int_0^{p_F} \frac{p^2c^2}{\epsilon_p} g(p)\mathrm{d}p, \quad \text{where} \quad g(p)\mathrm{d}p = \frac{2V}{h^3}4\pi p^2 \mathrm{d}p. \tag{6.11}$$

If we introduce dimensionless momenta $x = p/m_e c$, we obtain

$$P = \frac{8\pi m_e^4 c^5}{3h^3}\int_0^{x_F} \frac{x^4}{(1+x^2)^{1/2}}\mathrm{d}x. \tag{6.12}$$

The upper limit of the integral is the dimensionless Fermi momentum $x_F$, which, according to Eq. (2.27), is given by

$$x_F = \frac{p_F}{m_e c} = \left[\frac{3n_e}{8\pi}\right]^{1/3} \frac{h}{m_e c} = \left[\frac{3Y_e \rho_c}{8\pi m_{\rm H}}\right]^{1/3} \frac{h}{m_e c}. \tag{6.13}$$

Some integration and a bit of tidying-up leads to the following expression for the pressure:

$$P = K_{UR} n_e^{4/3}\, I(x_F), \tag{6.14}$$

where

$$I(x) = \frac{3}{2x^4}\left[x(1+x^2)^{1/2}\left(\frac{2x^2}{3} - 1\right) + \ln[x + (1+x^2)^{1/2}]\right]. \tag{6.15}$$

Because we have used Eq. (6.10), the exact relation between energy and momentum, Eq. (6.14) is the pressure due to an ideal degenerate electron gas of any density. This pressure is expressed in of terms of the dimensionless Fermi momentum $x_F$, which, according to Eq. (6.13), depends on the density. At high density the Fermi momentum is large and $x_F >> 1$. In this case the integral $I(x_F)$ tends to 1 and Eq. (6.14) gives a pressure in agreement with Eq. (6.6), the pressure of a gas of ultra-relativistic, degenerate electrons. At low density the Fermi momentum is small and $x_F << 1$. The integral $I(x_F)$ now tends to $4x_F/5$ and Eq. (6.14) gives a pressure in agreement with Eq. (6.2), the pressure in a non-relativistic, degenerate gas.

We can now consider the hydrostatic equilibrium of a star supported by electrons which form an ideal degenerate gas of any density. If we equate the pressure given by Eq. (6.14) to the pressure needed to support a star of mass $M$, we obtain

$$K_{UR}\left[\frac{Y_e \rho_c}{m_{\rm H}}\right]^{4/3} I(x_F) \approx \left[\frac{\pi}{36}\right]^{1/3} G M^{2/3} \rho_c^{4/3}, \tag{6.16}$$

which can be rearranged to give the following expression for the mass of the star:

$$M \approx [I(x_F)]^{3/2} M_{CH}, \tag{6.17}$$

where the Chandrasekhar mass $M_{CH}$ is given by Eq. (6.9).

Equation (6.17) gives the mass of a white dwarf in terms of $x_F$ which in turn depends on the central density $\rho_c$ via Eq. (6.13). Conversely, it gives the central density of a white dwarf with mass $M$. The results of an elementary calculation based on Eqs. (6.17) and (6.13), with $Y_e = 0.5$, are shown in Fig. 6.1. As expected, the central density increases as the mass of the white dwarf increases. Initially, the increase is in accord with Eq. (6.4), which is valid when the degenerate electrons

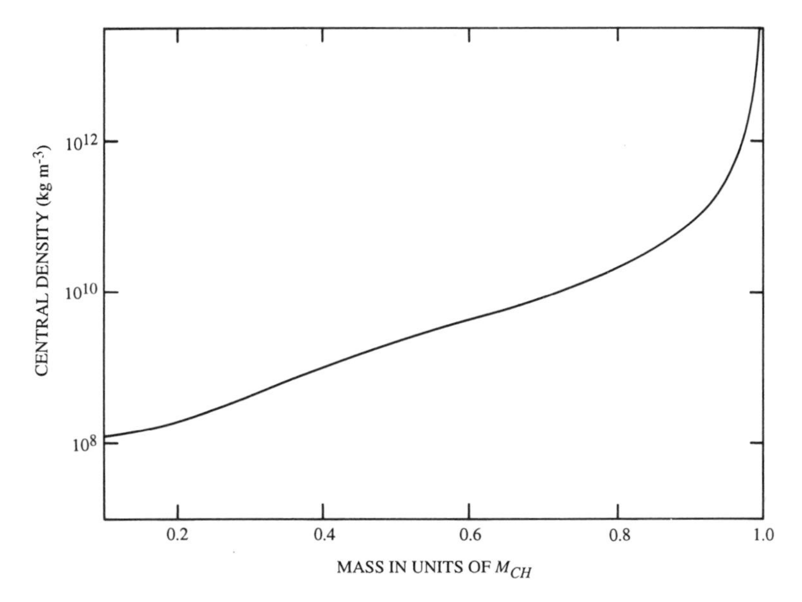


Fig. 6.1 The density at the centre of a white dwarf of mass $M$ supported by the pressure of an ideal gas of degenerate electrons. Note the density tends to infinity as the mass approaches the Chandrasekhar mass $M_{CH}$.

are non-relativistic. The density then increases more rapidly as the electrons become relativistic. Finally, as the mass approaches the Chandrasekhar mass, the electrons become ultra-relativistic and the density approaches infinity.

Our estimate, Eq. (6.9), for the magnitude of the Chandrasekhar mass is based upon the approximate relation between the central density and pressure given by Eq. (5.33). A more accurate estimate can be made if we use a polytrope model in which the relation between the density and pressure throughout the star is given by $P(r) \propto [\rho(r)]^{4/3}$, a relation which is consistent with Eq. (6.6), the pressure of a gas of ultra-relativistic, degenerate electrons. In this case the numerical factor in Eq. (5.33), which equals 0.44, is replaced by 0.36 and the value of the Chandrasekhar mass is predicted to be

$$M_{CH} \approx 3.1\, Y_e^2\, M_\star = 5.8\, Y_e^2\, M_\odot. \tag{6.18}$$

In most white dwarfs there are about two nucleons per electron and $Y_e \approx 0.5$ which, when substituted into Eq. (6.18), gives a Chandrasekhar mass of about 1.4 $M_\odot$.

Chandrasekhar first deduced that there is a maximum value for the mass of a white dwarf in 1931. It was a momentous discovery with a profound implication which he emphasized 1934 in the following way:

*The life history of a star of small mass must be essentially different from the life history of a star of large mass. For a star of small mass the natural white-dwarf stage is an initial step towards complete extinction. A star of large mass cannot pass into the white-dwarf stage and one is left speculating on the other possibilities.*

Later in the chapter we shall see that this speculation has led to the conclusion that the other possible end-points of stellar evolution are neutron stars and black holes. At this stage, we merely note that a star with a mass above the Chandrasekhar limit cannot form a stable white dwarf and that this instability can be traced to the fact that the degenerate electrons in the star are ultra-relativistic. This result is related to the general result discussed in Section 1.2 of Chapter 1; namely, that hydrostatic equilibrium becomes precarious for any star supported by a gas of ultra-relativistic particles.

### Mass and radius

According to Eq. (6.4), and more generally Fig. 6.1, the density of a white dwarf is a rapidly increasing function of its mass. This implies that the size of a white dwarf decreases with mass.

To explore the connection between the mass and radius of a white dwarf we need a model for the density distribution. If the degenerate electrons are predominantly non-relativistic, the structure of the star is similar to a polytrope model with $P \propto \rho^{5/3}$, in which case it can be shown that the average density is $\rho_c/6$. This, together with Eq. (6.4), implies that the average density of a white dwarf of mass $M$ is approximately

$$\langle \rho \rangle \approx \frac{0.51}{Y_e^5} \left[ \frac{M}{M_\star} \right]^2 \frac{m_\text{H}}{(h/m_e c)^3}, \tag{6.19}$$

and that the radius is

$$R = \left[ \frac{3M}{4\pi \langle \rho \rangle} \right]^{1/3} \approx 0.77\, Y_e^{5/3} \left[ \frac{M_\star}{M} \right]^{1/3} \alpha_G^{-1/2} \frac{h}{m_e c}. \tag{6.20}$$

In obtaining this last equation we have made use of the definition (6.5) of $M_\star$. Note that the characteristic size of a white dwarf is primarily determined by the fundamental constant $\alpha_G = 5.9 \times 10^{-39}$ and the electron Compton wavelength, $h/m_e c = 2.4 \times 10^{-12}$ m. This characteristic size is

$$\alpha_G^{-1/2} \frac{h}{m_e c} \approx 3 \times 10^7 \text{ m}. \tag{6.21}$$

We also note that the characteristic density is

$$\frac{m_{\rm H}}{(h/m_e c)^3} \approx 1 \times 10^8 \text{ kg m}^{-3}. \tag{6.22}$$

Further, if we use the sun as standard for mass and size, we find that a white dwarf with $Y_e = 0.5$ has a radius approximately given by

$$R \approx \frac{R_\odot}{74} \left[\frac{M_\odot}{M}\right]^{1/3}. \tag{6.23}$$

As expected, the radius of a white dwarf is a decreasing function of its mass. In deriving this mass–radius relation we assumed that the degenerate electrons were non-relativistic, and as such it is only applicable to low-mass white dwarfs. Nevertheless, it is in rough agreement with the limited observational data on the masses and radii of white dwarfs. Some of this data, which is limited because the mass can only be determined if the white dwarf is a member of a binary or triple system, is listed in Table 6.1. We note that the observed radii are comparable with the estimate given by Eq. (6.23) and that, as expected, the radius is a decreasing function of the mass.

TABLE 6.1 White dwarf masses and radii from optical observations. See Shapiro and Teukolsky (1983) for further details.

| White dwarf | Mass | Radius |
|---|---|---|
| Sirius B | $(1.053 \pm 0.028)\, M_\odot$ | $(0.0074 \pm 0.0006)\, R_\odot$ |
| 40 Eri B | $(0.48 \pm 0.02)\, M_\odot$ | $(0.0124 \pm 0.0005)\, R_\odot$ |
| Stein 2051 | $(0.50 \pm 0.05)$ or $(0.72 \pm .08)\, M_\odot$ | $(0.0115 \pm 0.0012)\, R_\odot$ |

The mass–radius relation (6.23) can be used to relate the luminosity of a white dwarf to its mass. We recall from Chapter 1 that the luminosity of a star depends upon its effective surface temperature $T_E$ and radius via the relation (1.43):

$$L = 4\pi R^2 \sigma T_E^4. \tag{6.24}$$

The mass–radius relation (6.23) then implies that a white dwarf of mass $M$ has a luminosity given by

$$L \approx \frac{1}{74^2} \left[\frac{M_\odot}{M}\right]^{2/3} \left[\frac{T_E}{6000}\right]^4 L_\odot. \tag{6.25}$$

For example, a white dwarf with $M = 0.4\, M_\odot$ and $T_E = 10^4$ K has a luminosity of about $3 \times 10^{-3}\, L_\odot$.

We saw in Section 3.4 of Chapter 3 that the rate of cooling of a white dwarf is largely determined by radiative diffusion through an outer, insulating layer which surrounds a largely isothermal, degenerate interior. Equation (6.25) shows that, as a white dwarf of a given mass cools, its declining luminosity and surface temperature are such that $L$ is proportional to $T_E^4$. This implies that a white dwarf cools along a specific line in the Hertzsprung–Russell diagram, as shown in Fig. 6.2. Moreover, because the position of the line of cooling is determined by the mass of the white dwarf and because all white dwarf masses lie in a narrow range, all white dwarfs are expected to occupy a narrow strip on the Hertzsprung–Russell diagram. The narrow mass range for white dwarfs arises from a precise upper limit and a less precise lower limit. Clearly, the mass cannot exceed the Chandrasekhar limit of about 1.4 $M_\odot$. Further the finite age of the universe implies that the mass of any observed white dwarf cannot be too low: this is the case because any observed white dwarf must have evolved from a main sequence star, and this star will evolve very slowly if its mass is low. With a universe only 10 to 20 billion years old, there has only been enough time for the evolution and emergence of white dwarfs with masses larger than 0.25 $M_\odot$, or thereabouts.

Finally, the mass–radius relation (6.23) can be used to estimate the strength of gravity on the surface of a white dwarf and to understand the gravitational red shift of radiation escaping from its surface.

We expect the acceleration due to gravity on a white dwarf to be very large by terrestrial or even solar standards. Using Eq. (6.23), we find that the acceleration on a white dwarf of mass $M$ is

$$g = \frac{GM}{R^2} \approx 74^2 \left[\frac{M}{M_\odot}\right]^{5/3} \frac{GM_\odot}{R_\odot^2}. \tag{6.26}$$

The numerical value of this acceleration on the surface of a star of mass 0.4 $M_\odot$ is approximately $4 \times 10^5$ m s$^{-2}$.

The gravitational red shift of radiation escaping from the surface is determined by the magnitude of the gravitational potential energy on its surface. According to General Relativity, the fractional change in the wavelength of the radiation is

$$\frac{\Delta\lambda}{\lambda} = \left[1 - \frac{2GM}{Rc^2}\right]^{-1/2} - 1 \approx \frac{GM}{Rc^2}. \tag{6.27}$$

One naive way of understanding this result is to consider that a photon of frequency $\nu$ has an effective mass $m = h\nu/c^2$ and a total energy of $h\nu - GmM/R$ on the surface. As the photon escapes, the gravitational potential energy increases and the frequency decreases in order to conserve energy. The change in frequency on escape

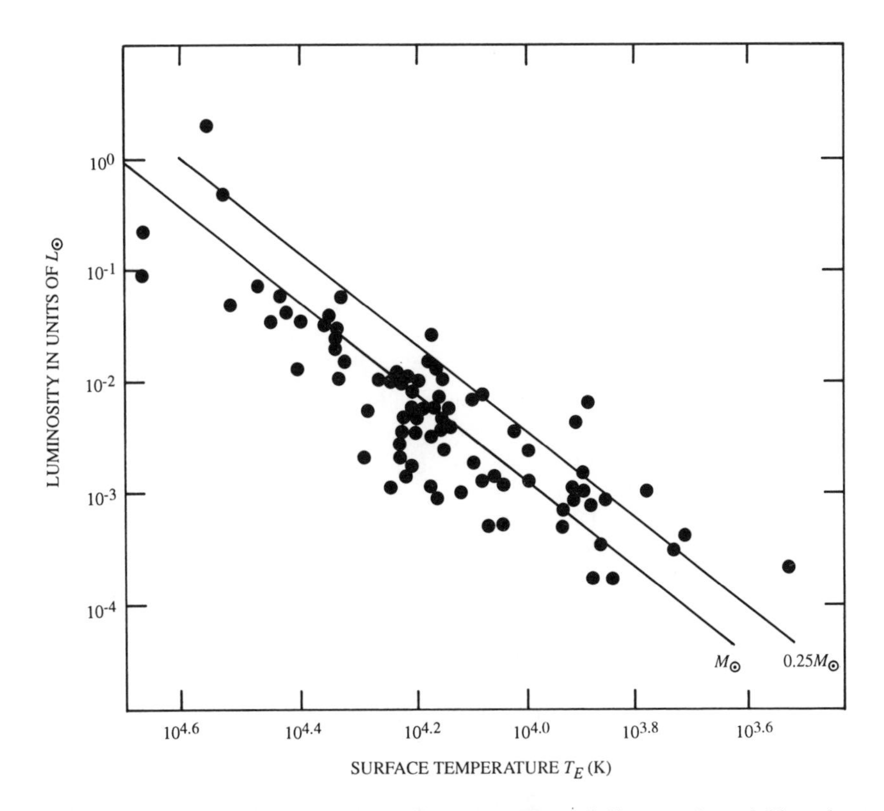


Fig. 6.2 Positions of white dwarfs on a Hertzsprung–Russell diagram. A model based on a star supported by a gas of non-relativistic, degenerate electrons leads to a relation between luminosity and surface temperature given by Eq. (6.25). The diagonal lines illustrate this relation for stars of mass $M= 0.25\,M_\odot$ and $M = M_\odot$. In fact, relativistic effects are important in massive white dwarfs, and their inclusion would reduce the predicted radius of the star with $M = M_\odot$ and the diagonal line would be shifted downwards. The observed positions of the white dwarfs in this diagram correspond to data compiled by Sweeney; see Shapiro and Teukolsky (1983) for further information.

is then $\Delta\nu = -GmM/Rh = -GM\nu/Rc^2$. Using the mass-radius relation (6.23), we find that the gravitational red shift for a white dwarf of mass $M$ is approximately

$$\frac{\Delta\lambda}{\lambda} \approx 74 \left[\frac{M}{M_\odot}\right]^{4/3} \frac{GM_\odot}{R_\odot c^2}. \tag{6.28}$$

For example, the observed red shift for 40 Eri B is $\Delta\lambda/\lambda = (7.97 \pm 0.43) \times 10^{-5}$. This white dwarf has a mass of $0.48\,M_\odot$, and the red shift expected from Eq. (6.28) is $\Delta\lambda/\lambda \approx 6 \times 10^{-5}$.

## 6.2 COLLAPSE OF A STELLAR CORE

A star with a mass greater than 11 $M_\odot$, or thereabouts, is expected to evolve through all the stages of nuclear burning. As outlined in Chapter 4, the process begins with hydrogen burning at about $2 \times 10^7$ K and proceeds at successively higher temperatures through helium, carbon, neon, oxygen and silicon burning. Silicon burning at about $3 \times 10^9$ K leads to a star with a central core of iron surrounded by concentric shells containing silicon, oxygen, neon, carbon, helium and hydrogen. Because energy cannot be released by the thermonuclear fusion of iron, the central core contracts. Initially, this contraction can be controlled by the pressure of the dense gas of degenerate electrons in the core. But as silicon burning in the surrounding shell deposits more iron onto the central core, the degenerate electrons in the core become increasingly relativistic. When the core mass reaches the Chandrasekhar limit of about 1.4 $M_\odot$, the electrons become ultra-relativistic and they are no longer able to support the core. At this stage the stellar core is on the brink of a catastrophe.

### The onset of collapse

When a body contracts under gravity, gravitational energy is converted into internal energy. If this leads to the activation of exothermic nuclear fusion, the internal kinetic energy increases, the pressure rises and the contraction is opposed. The opposite happens if an energy absorbing process is activated: kinetic energy is absorbed, the effectiveness of the pressure is diminished and gravitational contraction turns into gravitational collapse. Thus, a stellar boiler prevents gravitational contraction but a stellar refrigerator can trigger an uncontrolled collapse.

There are two energy absorbing processes, two possible refrigerators, which could drive the iron core of a star into an uncontrolled collapse. They are the photodisintegration of atomic nuclei and the capture of electrons via inverse beta decay. In the former, kinetic energy is used to unbind atomic nuclei, and in the latter, kinetic energy of degenerate electrons is converted into the kinetic energy of electron neutrinos which escape from the core. These energy-absorbing processes are so effective that the collapse of the stellar core is almost unopposed. Indeed, the core can collapse almost freely under gravity. According to Eq. (1.4), the time scale for such a collapse depends solely on the density of the core when the collapse is triggered. This density is expected to be around $10^{12}$ kg m$^{-3}$, and the free-fall collapse time of the core is remarkably short:

$$t_{FF} = \left[\frac{3\pi}{32G\rho}\right]^{1/2} \approx 1 \text{ millisecond.} \tag{6.29}$$

We shall briefly consider two energy-absorbing processes which could bring about a catastrophic collapse of this kind.

### Nuclear photodisintegration

As the stellar core contracts, the temperature increases and eventually a stage is reached when thermal photons are energetic enough to photodisintegrate iron nuclei; tightly bound iron nuclei are broken up into less tightly bound nuclei, and energy is absorbed. In a realistic calculation the whole range of possible nuclei should be considered. However, a useful insight can be obtained if we assume that a stage is reached when tightly bound $^{56}$Fe nuclei coexist with neutrons and tightly bound $^{4}$He nuclei, a coexistence governed by the reactions

$$\gamma + {}^{56}\mathrm{Fe} \rightleftharpoons 13\,{}^{4}\mathrm{He} + 4n. \tag{6.30}$$

In the following we shall label the reacting particle by its mass number $A$.

The photodisintegration of $^{56}$Fe is an endothermic reaction absorbing

$$Q = (13m_4 + 4m_1 - m_{56})c^2 = 124.4 \text{ MeV}. \tag{6.31}$$

Thus, one kilogram of iron could absorb $2 \times 10^{14}$ J, the energy equivalent of 50 kilotons of TNT.

The fraction of iron dissociated at a given temperature and density can be estimated by using the same techniques we used when we considered atomic ionization in Chapter 2, and helium and silicon burning in Chapter 4. We assume that the reactions (6.30) lead to thermodynamic equilibrium, so that the chemical potentials of the reacting particles satisfy the equation

$$\mu_{56} = 13\mu_4 + 4\mu_1. \tag{6.32}$$

According to Eq. (2.21) or Eq. (4.51), the chemical potential for a particle of mass number $A$ is

$$\mu_A = m_A c^2 - kT \ln\left[\frac{g_A n_{QA}}{n_A}\right], \tag{6.33}$$

and the quantum concentration $n_{QA}$ is

$$n_{QA} = \left[\frac{2\pi m_A kT}{h^2}\right]^{3/2}. \tag{6.34}$$

It follows that the equilibrium concentrations of the reacting particles are given by

$$\frac{(n_4)^{13}(n_1)^4}{n_{56}} = \frac{(g_4)^{13}(g_1)^4}{g_{56}}\frac{(n_{Q4})^{13}(n_{Q1})^4}{n_{Q56}} \exp[-Q/kT]. \tag{6.35}$$

The statistical factors $g_A$ depend on the angular momentum of the particle. For the spin-half neutron, $g_1 = 2$. For the $^4$He and $^{56}$Fe nuclei we can take $g_4 = 1$ and $g_{56} = 1$, if we assume that all the $^4$He and $^{56}$Fe nuclei are mostly in their spin-zero ground states. It is then easy to show that Eq. (6.35) implies that about three-quarters of the iron is dissociated when the density and temperature of the core reach $\rho = 10^{12}$ kg m$^{-3}$ and $T = 10^{10}$ K.

At higher temperatures the $^4$He nuclei are also expected to dissociate via the reactions

$$\gamma + {}^4\text{He} \rightleftharpoons 2p + 2n. \tag{6.36}$$

Again, it is straightforward to find the degree of dissociation by equating chemical potentials; see problem 6.2 at the end of the chapter.

It is easy to estimate the total energy that could be absorbed by these photodisintegration processes. Bearing in mind that the collapsing iron core has a mass comparable with the Chandrasekhar mass of 1.4 $M_\odot$, $4 \times 10^{44}$ J are absorbed by the photodisintegration of $^{56}$Fe nuclei and a further $1 \times 10^{45}$ J by a subsequent photodisintegration of $^4$He nuclei. Thus, the total energy that could be absorbed by the photodisintegration of the iron core to neutrons and protons is approximately

$$E_{photo} \approx 1.4 \times 10^{45} \text{ J}. \tag{6.37}$$

This is a substantial energy, equivalent to the energy radiated by the sun over a period of 10 billion years. There is no doubt that the absorption of an energy of this magnitude could trigger an uncontrolled collapse of the stellar core.

### Electron capture

In normal circumstances a neutron is an unstable particle with a half-life of 10.25 minutes. It decays into a proton, an electron and a neutrino via the beta decay

$$n \rightarrow p + e^- + \overline{\nu}_e. \tag{6.38}$$

The electron and the neutrino produced in this decay have a combined energy of 1.3 MeV, an energy equal to the mass-energy difference of a neutron and a proton. Thus, electrons with energies up to 1.3 MeV are produced when neutrons decay. It follows that neutrons will not be able to decay, if electrons with these energies cannot be produced. This can be achieved by immersing the neutrons in a dense gas of degenerate electrons so that all the electron states with an energy up to 1.3 MeV are fully occupied in accordance with the Pauli exclusion principle. The required density of the electron gas can be found by recalling that the maximum momentum of a electron in a degenerate gas, the Fermi momentum, is

$$p_F = h\left[\frac{3n_e}{8\pi}\right]^{1/3}, \tag{6.39}$$

and that the maximum energy, the Fermi energy, is given by

$$\epsilon_F^2 = p_F^2c^2 + m_e^2c^4. \tag{6.40}$$

Furthermore, if the gas is denser than this critical density, electrons with an energy greater than 1.3 MeV exist and they may be captured by protons to form neutrons by the inverse beta decay process,

$$e^- + p \rightarrow n + \nu_e. \tag{6.41}$$

This conversion of protons to neutrons is often called neutronization.

In practice, the protons in the core of an evolved, massive star are not free but bound in atomic nuclei. Nevertheless, they can still capture energetic electrons to form neutrons, and in so doing they produce nuclei which are increasingly rich in neutrons. Neutronization begins in the stellar core when the main constituent, $^{56}$Fe, can undergo the inverse beta decay,

$$e^- + {}^{56}\text{Fe} \rightarrow {}^{56}\text{Mn} + \nu_e. \tag{6.42}$$

This will be energetically possible when the density of the contracting iron core reaches $1.1 \times 10^{12}$ kg m$^{-3}$; at this density the Fermi energy of the electrons equals the threshold energy of $m_ec^2 + 3.7$ MeV needed for the inverse beta decay of $^{56}$Fe. Normally, a $^{56}$Mn nucleus beta decays to $^{56}$Fe with half-life of 2.6 hours, but in the stellar core it captures an electron from the dense degenerate gas to form a $^{56}$Cr nucleus. This in turn is capable of capturing an electron when the density reaches $1.5 \times 10^{13}$ kg m$^{-3}$.

Electron capture by inverse beta decay on nuclei in the stellar core becomes very rapid when the density exceeds $10^{14}$ kg m$^{-3}$. The neutrinos produced interact very weakly with matter and carry away the energy originally stored by degenerate electrons. As the pressure generated by these electrons disappears, the stellar core collapses rapidly.

It is easy to estimate the possible energy loss due to electron capture in the stellar core. First, we note that an iron core with a mass equal to the Chandrasekhar mass contains about $10^{57}$ electrons which could produce $10^{57}$ electron neutrinos. Second, we assume that the average energy of a captured electron is around 10 MeV; this corresponds to the average energy of an degenerate electron when the density of the core is $2 \times 10^{13}$ kg m$^{-3}$. Thus, the total energy that could be lost by electron capture is

$$E_{cap} \approx 10^{57} \times (10 \times 1.6 \times 10^{-13}) = 1.6 \times 10^{45}\ \text{J}. \tag{6.43}$$

This energy is carried away from the star by a burst of electron neutrinos. If the neutrinos escaped freely, the duration of this burst would be comparable with the millisecond time scale for the free fall of the core under gravity given by Eq. (6.29). However, many of the neutrinos interact with the dense matter formed by the collapsed core. Indeed, theoretical calculations indicate that the neutrino mean free path becomes comparable with the size of the core when the core radius is a few kilometres and the density is $10^{14}$ kg m$^{-3}$. Neutrinos, which earlier streamed out of the imploding core, are now trapped in the implosion. Because of this, most of the electron neutrinos formed by electron capture will be trapped for a few seconds before they diffuse out of the collapsed core.

## The aftermath

We have seen that electron capture and/or photodisintegration can trigger the collapse of the iron core of a massive star. The collapse is rapid and almost unopposed until a density comparable to the density of nuclear matter is reached. This density can be determined from the well-known formula for the radius of a nucleus containing $A$ nucleons: a radius given by

$$R = r_0 A^{1/3} \quad \text{where} \quad r_0 = 1.2 \times 10^{-15} \text{ m}, \tag{6.44}$$

implies a nuclear density of

$$\rho_{nuc} = \frac{3Am_N}{4\pi R^3} = \frac{3m_N}{4\pi r_0^3} = 2.3 \times 10^{17} \text{ kg m}^{-3}, \tag{6.45}$$

where $m_N$ is the nucleon mass. Clearly, neutron degeneracy and nuclear forces in the neutron-rich core will begin to be important when the nuclear density $\rho_{nuc}$ is reached. Moreover, nuclear forces are expected to resist compression and bring the collapse to a halt when the core becomes two or three times more dense than normal nuclear matter. The core is expected to rebound strongly and set up a shock wave that travels through the material that is falling towards the centre. Theoretical calculations suggest that this shock wave may be able to reverse the inward fall of stellar material surrounding the core and produce an outward expulsion, a supernova.

Supernovae are very energetic explosions: the observed kinetic energy of the debris is typically $10^{44}$ J and the optical energy output during the year following the explosion is of the order of $10^{42}$ J. These explosions are primarily classified according to their optical spectra, with particular attention to the presence or absence of spectral lines associated with hydrogen. However, this spectral classification bears little relation to the underlying cause of the explosion. Two possible causes could yield the required energy, either the disruption of a star by a thermonuclear detonation or the collapse of the iron core of a massive star.

In passing, we shall briefly describe a possible scenario for a supernova due to a thermonuclear detonation of a star. This scenario involves a carbon–oxygen white dwarf which can increase its mass by drawing mass from a nearby companion star. When the mass of the white dwarf exceeds the Chandrasekhar limit of 1.4 $M_\odot$, it contracts and ignites the thermonuclear fusion of the hitherto quiescent carbon and oxygen. Because this material is degenerate, the fusion-control mechanism discussed at the end of Section 1.4 is not operative. We recall that an energy release in a star leads to an expansion and an accompanying decrease in the internal kinetic energy. The latter normally implies a temperature decrease and a reduction of the fusion rate. In degenerate matter, however, the decrease in the internal kinetic energy lowers the energy of the degenerate electrons and has little effect on the temperature so that the rate of fusion is uncontrolled. Thus, the sudden ignition of thermonuclear fusion in a white dwarf creates a star-sized fusion bomb. The white dwarf could explode more or less as a whole, leaving no residual core behind.

We now return to the collapse of an iron core of a massive star. The physics of the collapse, the rebound and associated shock wave is very complicated. As we shall see in a moment, a gigantic amount of gravitational energy is released. But it is not certain how energy and momentum is transferred to the outer layers of the star, and it is by no means certain that core collapse is always accompanied by a supernova. Nevertheless, the collapse is expected to leave a core residue, either a neutron star or an over-weight neutron star which collapses to form a black hole. We shall take a closer look at neutron stars and briefly comment on black holes later in this chapter. At this stage we shall concentrate on the energy of formation of a neutron star, the energy that must be released when the neutron star is formed. In so doing, we shall show that the fireworks of any accompanying supernova is, if assessed in terms of energy, an insignificant side-show.

The energy of formation of a neutron star is largely determined by the change in the gravitational binding caused by core collapse. Just before collapse we have a core with a mass comparable to the sun and a radius of about 1000 km. After the collapse we have a neutron star with a similar mass but with a radius of about 10 km. The initial gravitational binding is negligible and the gravitational energy released in the collapse is simply the gravitational binding of the neutron star. For a neutron star of mass $M$ and radius $R$, this binding energy is approximately given by

$$E_B \approx \frac{GM^2}{R} = 3 \times 10^{46} \left[\frac{M}{M_\odot}\right]^2 \left[\frac{10 \text{ km}}{R}\right] \text{ J.} \tag{6.46}$$

We emphasize that the energy of formation implied by Eq. (6.46) is an order of magnitude greater than the energy needed to photodisintegrate the iron nuclei in the core; see Eq. (6.37). It is also an order of magnitude greater than the energy lost by electron capture; see Eq. (6.43). Moreover, it is much larger than the energy associated with the kinetic and visible effects of any supernova which may be triggered by the collapse; as mentioned earlier, the typical kinetic energy of the debris of a supernova is only $10^{44}$ J and the optical energy output during the year following the explosion is only of the order of $10^{42}$ J. Thus, we still have to account for about 90% of the energy released when a neutron star is formed by core collapse. There must be an important intermediate stage before the formation of a compact neutron star, a stage characterized by an energy loss of about $3 \times 10^{46}$ J, the typical binding energy of a neutron star.

This intermediate stage is thought to be the formation of a hot, bloated neutron star which then cools and contracts by emitting neutrinos. This bloated neutron star is, to a first approximation, a dense plasma of neutrons, protons, nuclei, electrons, photons and neutrinos held together by gravity; the typical temperature and density are of the order of $10^{11}$ K and $10^{14}$ kg m$^{-3}$. The plasma is almost completely opaque to photons and little energy escapes by electromagnetic radiation. Instead, cooling occurs by the emission of neutrinos. These weakly interacting particles can travel several metres in the plasma before interaction. They escape from the hot neutron star by a random walk process similar to that discussed in Section 1.4. If their mean free path is $\bar{l}$, they will interact about $R^2/\bar{l}^2$ times before they escape from the surface of a star with radius $R$. Hence the escape time will be of the order of $R^2/\bar{l}c$.

Only a fraction of the escaping neutrinos arise from electron capture. The hot, bloated neutron star is so hot that neutrino–antineutrino pairs are copiously produced. The simplest production mechanism, the annihilation of an electron–positron pair, was briefly discussed in Section 2.6, but other mechanisms, such as plasmon decay, photoneutrino production and neutrino bremsstrahlung, are also thought to be important.

Three types of neutrino–antineutrino pairs can be produced. They are denoted by $\nu_e$, $\overline{\nu}_e$ and $\nu_\mu$, $\overline{\nu}_\mu$ and $\nu_\tau$, $\overline{\nu}_\tau$. The $\nu_e$ neutrino is associated with the most familiar lepton of all, the electron. The neutrinos $\nu_\mu$ and $\nu_\tau$ are associated with massive, unstable charged particles called muons and tauons, particles very similar to electrons but with masses 106 MeV$/c^2$ and 1784 MeV$/c^2$. The antineutrinos $\overline{\nu}_e$, $\overline{\nu}_\mu$ and $\overline{\nu}_\tau$ are associated with the antielectron or positron, the antimuon and the antitauon, respectively. The various types of neutrinos and antineutrinos are very similar. They are all weakly interacting fermions with a mass which is either zero or very small. However, they do differ from each other, and this difference can be illustrated by how they interact. For example, a weak interaction could transform a $\nu_e$ into an electron, but not into a muon or tauon.

Because the masses of these neutrinos are either zero or very small, all three types of neutrino–antineutrino pairs are produced by thermal processes in the hot neutron star. In all, six kinds of weakly interacting particles are formed: $\nu_e$ and $\overline{\nu}_e$, $\nu_\mu$ and $\overline{\nu}_\mu$, and $\nu_\tau$ and $\overline{\nu}_\tau$. These particles carry away the bulk of the binding energy of the neutron star. Moreover, each of the six kinds of weakly interacting particle is expected to carry away about one-sixth of this energy. Thus, each kind of neutrino carries away about $0.5 \times 10^{46}$ J if the binding energy of the neutron star is $3 \times 10^{46}$ J. The time scale for this cooling process is the time needed for a neutrino to diffuse to the surface of the neutron star. As mentioned earlier, this is of the order of $R^2/\bar{l}c$.

To summarize, the authentic signature of the aftermath of core collapse is not a supernova, but an intense pulse of neutrinos. An observation of the energy of these neutrinos and the duration of the pulse would reveal how gravitational collapse is shaped by weak interaction processes.

The detection of neutrinos from core collapse is a formidable problem. If the neutrinos can escape from a hot, neutron star and penetrate the outer layers of the collapsing star, they are more than likely to pass through any detection apparatus. Similar problems are met in the detection of neutrinos from the sun, as discussed in Section 4.2 of Chapter 4. But the techniques for the detection of neutrinos from core collapse are different, and somewhat easier, because these neutrinos are more energetic than solar neutrinos. These techniques were successfully demonstrated for the first time on 23 February 1987, when two massive underground detectors, the Kamiokande II (KII) detector in Japan and the Irvine–Michigan–Brookhaven (IMB) detector in the US, detected neutrinos from the supernova SN1987A.

The KII and IMB detectors are similar in design, consisting of large volumes of ultra-pure water surrounded by thousands of photomultiplier tubes. The neutrino burst from SN1987A was mainly detected via the reaction,

$$\overline{\nu}_e + p \rightarrow n + e^+. \tag{6.47}$$

If the positron recoils with a velocity greater than the phase velocity of light in the water, it emits Čerenkov radiation which can be detected by photomultiplier tubes surrounding the water. The $\overline{\nu}_e$ absorption reaction (6.47) is the most probable reaction involving neutrinos and antineutrinos from the supernova. Even so, less than one in $10^{15}$ of the $\overline{\nu}_e$'s from the supernova were detected. The data from the KII and IMB detectors is illustrated in Fig. 6.3.

First and foremost we note that only 20 neutrinos were detected by the KII and IMB detectors.

Second, the duration of the neutrino pulse, as illustrated in Fig. 6.3, is about 10 seconds. This time can be identified with $R^2/\bar{l}c$, the typical time needed for a neutrino with mean free path $\bar{l}$ to diffuse from the cooling neutron star of radius $R$. For example, a time of the right order of magnitude is obtained if $R = 100$ km and $\bar{l} = 10^{-4}\,R$.

Third, when account is taken of the efficiency of detection and of the distance of supernova SN1987A, about 50 kpc, the data in Fig. 6.3 is consistent with $\overline{\nu}_e$ radiation with a total energy between $0.3\times10^{46}$ to $0.5\times10^{46}$ J, an energy comparable with one sixth of the expected binding energy of a neutron star.

Finally, the observed energies of the detected neutrinos are consistent with the energy spectrum expected from a 'black-body' neutrino radiator at an effective temperature of $T_E \approx 5 \times 10^{10}$ K, some ten million times hotter than the effective

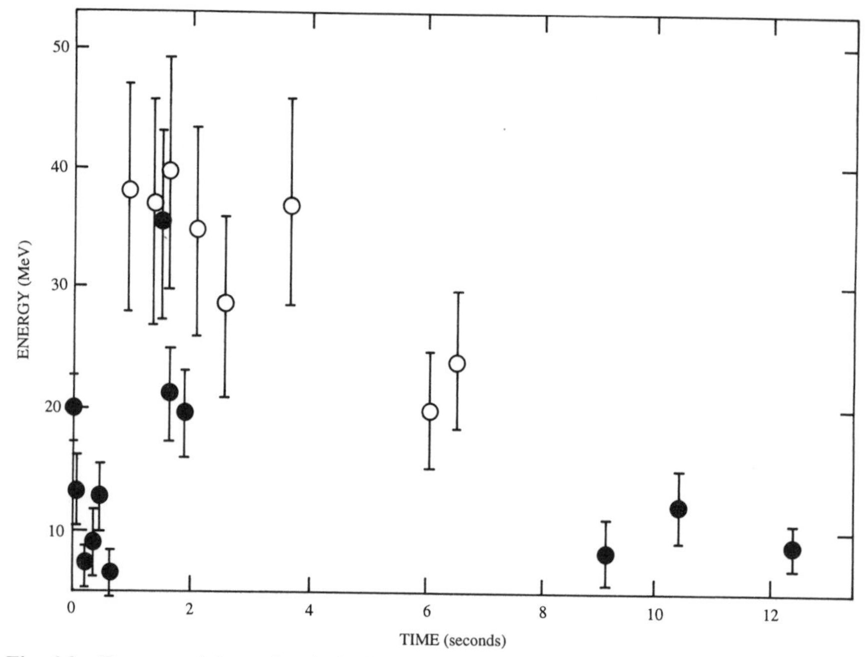


Fig. 6.3 Energy and time of arrival of neutrinos from the supernova SN1987A as registered by the Kamiokande II and IMB detectors. In all 20 neutrinos were detected and the duration of the neutrino pulse was about 10 seconds.

surface temperature of the sun.[1] Indeed, many of the qualitative arguments used in Chapter 1 to describe the diffusion of photons from the sun are applicable to the diffusion of neutrinos from a cooling neutron star. In particular, we can adapt Eq. (1.35) to find an approximate relation between the typical internal temperature $T_I$ and the effective surface temperature $T_E$ of the cooling neutron star,

$$T_E \approx \left[\frac{\bar{l}}{R}\right]^{1/4} T_I. \tag{6.48}$$

Hence, if the characteristic mean free path for neutrinos inside the cooling neutron star is $10^{-4}$ $R$, a value consistent with the observed duration of the neutrino pulse, then the typical internal temperature is between $10^{11}$ and $10^{12}$ K.

All in all, the detection of 20 neutrinos from SN1987A by the KII and IMB detectors gave credence to many of the theoretical expectations for the aftermath of the collapse of the iron core of a massive star. It was one of the most exciting

[1] In problem 6.3 at the end of the chapter you are asked to show that the average energy of neutrinos from such a radiator at temperature $T$ is 3.15 $kT$.

astrophysical events of the century. It established a new branch of astronomy: non-solar neutrino astronomy. Managers of scientific research should take careful note that both the KII and IMB detectors were designed and built for another purpose, the observation of proton decay.

## 6.3 NEUTRON STARS

A neutron star is born as a hot residue of the collapsed core of a massive star. The typical internal temperature is initially between $10^{11}$ to $10^{12}$ K. It rapidly cools by neutrino emission and is expected to reach a temperature of the order of $10^9$ K in a day and $10^8$ K in a 100 years. These are high temperatures by terrestrial and solar standards, but they are low by the standards set by the high densities of the matter inside a neutron star; the electrons, protons and above all neutrons inside a neutron star are degenerate and occupy the lowest possible states consistent with the Pauli exclusion principle. We begin by commenting on how the nature of the matter inside a neutron star depends upon its density.

### Matter inside neutron stars

In normal circumstances the most stable form of nuclear matter consists of nuclei near $^{56}$Fe in the periodic table. Less massive nuclei are less stable because they have a higher fraction of their nucleons near the surface, and more massive nuclei are less stable because of the increased importance of Coulomb repulsion between protons. The best deal, with the lowest binding energy per nucleon, is struck near $^{56}$Fe.

This deal is changed by the presence of relativistic electrons. As described in Section 6.2, degenerate electrons in a collapsed star are sufficiently energetic to induce inverse beta decay. Protons are converted to neutrons, and nuclei rich in neutrons are formed. Coulomb forces now have a reduced importance and neutron-rich nuclei, heavier than $^{56}$Fe, are energetically favoured. For example, $^{78}$Ni and $^{76}$Fe are thought to be the most stable nuclei in an electron gas when the density is around $10^{14}$ kg m$^{-3}$.

When the density exceeds $4 \times 10^{14}$ kg m$^{-3}$, a new phenomenon occurs called *neutron drip*. Neutrons drip from neutron-rich nuclei so that free neutrons, nuclei and electrons coexist in equilibrium. The equation of state for this form of matter is well understood for densities below $\rho_{nuc} = 2.3 \times 10^{17}$ kg m$^{-3}$, the density of normal nuclear matter. At higher densities nuclei begin to merge with each other and a dense gas of electrons, protons and neutrons is formed. The equation of state now strongly depends upon the interaction between nucleons, an interaction which is both complicated and uncertain. At higher densities, around $10^{18}$ kg m$^{-3}$, further complexities and uncertainties are introduced as it becomes energetically possible to produce pions, muons and hyperons. At higher densities still, the quark degrees of freedom are expected to play a role.

In order to gain some insight into why neutrons are the dominant constituent of neutron stars, we shall crudely neglect interactions and consider an ideal gas of degenerate electrons, protons and neutrons. At high densities, neutrons are present in this gas because their normal beta-decay mode, $n \rightarrow p + e^- + \overline{\nu}_e$, is blocked by the Pauli exclusion principle; the decay does not occur because it would involve the emission of either a proton or an electron into a state which is already fully occupied.

It is sufficient to consider the possible decay of one of the most energetic neutrons present, one with an energy equal to the neutron Fermi energy $\epsilon_F(n)$. Such a neutron cannot decay if the emitted proton and electron have energies below the Fermi energies for protons and electrons, $\epsilon_F(p)$ and $\epsilon_F(e)$. It follows that all the neutrons with energies up to $\epsilon_F(n)$ are stabilized by the Pauli principle if

$$\epsilon_F(n) < \epsilon_F(p) + \epsilon_F(e). \tag{6.49}$$

Conversely, neutrons can beta decay if

$$\epsilon_F(n) > \epsilon_F(p) + \epsilon_F(e). \tag{6.50}$$

In fact, the coexistence in equilibrium of neutrons, protons and electrons at zero temperature is characterized by

$$\epsilon_F(n) = \epsilon_F(p) + \epsilon_F(e). \tag{6.51}$$

This result can also be obtained by noting that the chemical potential of a Fermi gas at zero temperature is the Fermi energy. Thus, Eq. (6.51) is a relation between chemical potentials which characterizes the equilibrium established at zero temperature by the processes,

$$n \rightarrow p + e^- + \overline{\nu}_e \quad \text{and} \quad e^- + p \rightarrow n + \nu_e, \tag{6.52}$$

with the neutrinos playing no part because they escape.

The equilibrium concentrations of the neutrons, protons and electrons, $n_n$, $n_p$ and $n_e$, implied by Eq. (6.51) can be found by noting that the Fermi momentum of a particle is related to its concentration by Eq. (2.27), namely

$$p_F = \left[\frac{3n}{8\pi}\right]^{1/3} h. \tag{6.53}$$

When the density is of the order of $\rho_{nuc}$, the neutrons and the protons are approximately non-relativistic with Fermi energies and momenta related by

$$\epsilon_F(n) \approx m_n c^2 + \frac{p_F(n)^2}{2m_n} \quad \text{and} \quad \epsilon_F(p) \approx m_p c^2 + \frac{p_F(p)^2}{2m_p}. \tag{6.54}$$

The less massive electrons, however, are ultra-relativistic and the relation between the electron Fermi energy and momentum is

$$\epsilon_F(e) \approx p_F(e)c. \tag{6.55}$$

Bearing in mind that neutron-star matter is neutral with equal numbers of electrons and protons, we set $n_e = n_p$ and find the following relation between the numbers of neutrons and protons in the ideal gas at equilibrium:

$$\left[\frac{3n_p}{8\pi}\right]^{1/3} hc + \left[\frac{3n_p}{8\pi}\right]^{2/3} \frac{h^2}{2m_p} - \left[\frac{3n_n}{8\pi}\right]^{2/3} \frac{h^2}{2m_n} \approx m_n c^2 - m_p c^2. \tag{6.56}$$

Given the neutron–proton mass difference of 1.3 MeV$/c^2$, it is straightforward to find the relative numbers of neutrons and protons at any particular density. For example, at a typical neutron-star density of $\rho = 2 \times 10^{17}$ kg m$^{-3}$, we find $n_n \approx 1 \times 10^{44}$ m$^{-3}$ and $n_e = n_p \approx n_n/200$; i.e. one electron per 200 neutrons is enough to prevent neutron decay. We conclude that neutrons are the dominant constituent of neutron-star matter at densities of the order of $10^{17}$ kg m$^{-3}$.

### The size of neutron stars

We shall now take the simplistic, ideal degenerate gas model for the material inside a neutron star one step further, and investigate how the central density and radius of the star depend upon its mass. This can be done very simply by adapting the analysis of white dwarfs given in Section 6.1. This analysis assumed that a white dwarf is supported by the pressure of an ideal gas of degenerate electrons. We now assume that a neutron star is supported by the pressure of an ideal gas of degenerate neutrons.

Because neutrons are the dominant constituent of the star, the number density of neutrons is directly determined by the mass density. At the centre of the star

$$n_n \approx \frac{\rho_c}{m_n}. \tag{6.57}$$

The corresponding equation for a white dwarf is Eq. (6.1). The white-dwarf equations, Eqs. (6.2) to (6.4), can be modified so as to describe the hydrostatic equilibrium of a neutron star by changing the electron mass to the neutron mass and by setting $Y_e$ equal to one. We also ignore the difference between the mass of the hydrogen atom $m_{\rm H}$ and the mass of the neutron $m_n$; this difference is less than 0.1%. In this way, we deduce that a gas of degenerate, non-relativistic neutrons can support a neutron star of mass $M$ if the central density is

$$\rho_c \approx 3.1 \left[\frac{M}{M_\star}\right]^2 \frac{m_n}{(h/m_n c)^3}. \tag{6.58}$$

The fundamental stellar mass $M_\star$ is given by Eq. (6.5), but in the context of a neutron star it is best expressed in terms of the neutron mass as follows:

$$M_\star = \alpha_G^{-3/2} m_n = 1.85\ M_\odot. \tag{6.59}$$

The radius of the neutron star can be found by adapting Eq. (6.20). We find that

$$R \approx 0.77 \left[\frac{M_\star}{M}\right]^{1/3} \alpha_G^{-1/2} \frac{h}{m_n c}. \tag{6.60}$$

We note that the characteristic size of a neutron star is primarily determined by the dimensionless measure of the strength of gravity, $\alpha_G = 5.9 \times 10^{-39}$, and the Compton wavelength of the neutron, $h/m_n c = 1.3 \times 10^{-15}$ m. This characteristic size is

$$\alpha_G^{-1/2} \frac{h}{m_n c} \approx 17 \text{ km}, \tag{6.61}$$

which is about 2000 times smaller than the typical size of a white dwarf given by Eq. (6.21).

It is important to emphasize that the expression (6.60) for the radius of a neutron star is very approximate and rests on a number of assumptions of doubtful validity. In particular, the interactions between the neutrons cannot be neglected at neutron-star densities. Moreover, relativistic effects can be important. Indeed, because neutrons in a degenerate gas have momenta comparable with $m_n c$ when the density approaches $m_n/(h/m_n c)^3$, Eq. (6.58) implies that relativistic effects are only unimportant in neutron stars with masses much smaller than $M_\star$. In addition, the gravitational fields in neutron stars are very large and Einstein's theory of gravitation, and not Newton's, should really be used in establishing the condition of hydrostatic equilibrium. An indication of whether Newtonian gravitation is an adequate approximation is provided by the smallness of the ratio of the gravitational potential energy to the rest-mass energy of a particle on the surface of a neutron star. Using Eq. (6.60), we find that this ratio is

$$\frac{GM}{Rc^2} \approx 0.2 \left[\frac{M}{M_\star}\right]^{4/3}. \tag{6.62}$$

We conclude that the gravitational fields of a neutron star are only Newtonian if the mass is small compared with $M_\star$.

Despite these misgivings about the accuracy of Eq. (6.60), this equation for the radius of a neutron star can yield useful estimates of some important neutron-star properties.

### Gravitational binding energy of neutron stars

We saw in Section 6.2 that the gravitational binding energy of a neutron star is an important property. It is approximately equal to the energy emitted as neutrino radiation during the collapse of a stellar core. It is straightforward to estimate this binding energy. Using Eq. (6.60), or more directly Eq. (6.62), we find that the binding energy of a neutron star of mass $M$ is approximately

$$E_B \approx \frac{GM^2}{R} \approx 0.2 \left[\frac{M}{M_\star}\right]^{7/3} M_\star c^2 = \left[\frac{M}{M_\star}\right]^{7/3} 7 \times 10^{46}\ \mathrm{J}. \qquad (6.63)$$

We note that this estimate is compatible with that used when we considered the energy of formation of neutron stars in Section 6.2; see Eq. (6.46). We also note that the binding energy of a neutron star is only small compared with its rest mass energy if its mass is small compared with $M_\star$. This is yet another indication that relativistic effects are important in massive neutron stars.

### Rotating neutron stars and pulsars

The possibility of the existence of neutron stars was postulated very soon after Chadwick's discovery of the neutron in 1932. In 1934, Baade and Zwicky tentatively linked supernovae with the collapse of ordinary stars to neutron stars, and the first theoretical models for neutron stars were developed by Oppenheimer and Volkoff in 1939. However, there was surprisingly little astronomical and theoretical interest in neutron stars until the accidental observational discovery of pulsars by Hewish and Bell in 1967.

Pulsars emit pulses of radiation at short and remarkably regular intervals. Many pulsars have been observed with periods ranging from milliseconds to seconds. But the most famous pulsar is at the heart of the Crab Nebula, the remnant of a supernova which, according to Chinese historical records, occurred in 1054 AD. The Crab Pulsar has a period of 33 ms; it is also slowing down so that its period increases by a millisecond every 90 years.

The identification in the late 1960s of newly discovered pulsars with rotating neutron stars stimulated a renewed interest in the physics of neutron stars.

The principal argument identifying pulsars with neutron stars is based upon the shortness of pulsar periods. This argument can be understood by considering the maximum speed of rotation of a star. Bearing in mind that matter will be thrown off the star if it rotates too quickly, we can find the maximum angular frequency, and the corresponding minimum period, by equating the gravitational attraction at the surface of the star to the centrifugal force tending to throw matter off the star; this is the condition for weightlessness on the surface of the star. This condition leads to

$$\frac{GM}{R^2} = R\omega_{max}^2 \quad \text{and} \quad \tau_{min} = \frac{2\pi}{\omega_{max}} = 2\pi \left[\frac{R^3}{GM}\right]^{1/2}. \tag{6.64}$$

If the radius is given by Eq. (6.60), we find that the minimum period of rotation of a neutron star of mass $M$ is

$$\tau_{min} \approx 11 \left[\frac{M_\star}{M}\right] \alpha_G^{-1/2} \frac{h}{m_n c^2} = 0.6 \left[\frac{M_\star}{M}\right] \text{ms.} \tag{6.65}$$

This implies that a neutron star with the mass of the sun could rotate with a period as short as a millisecond without ripping itself apart. We note from Eq. (6.64) that the possibility of rapid rotation is a direct consequence of the high density of neutron stars; less dense objects, such as white dwarfs, could not rotate as quickly. Hence, the pulsar at the heart of the Crab Nebula, whose period is 33 ms, cannot be a rotating white dwarf. It is almost certainly a rotating neutron star.

The next property to be considered is the moment of inertia of a neutron star. The moment of inertia of a sphere of uniform density is

$$I = \frac{2}{5} MR^2. \tag{6.66}$$

For a neutron star with radius given by Eq. (6.60), we find

$$I \approx 0.24 \left[\frac{M}{M_\star}\right]^{1/3} \alpha_G^{-5/2} m_n \left[\frac{h}{m_n c}\right]^2 = \left[\frac{M}{M_\star}\right]^{1/3} 2.5 \times 10^{38} \text{ kg m}^2. \tag{6.67}$$

This estimate for the moment of inertia of a neutron star can be used to provide additional evidence in favour of the identification of the Crab Pulsar with a rotating neutron star.

As mentioned earlier, the Crab Pulsar is slowing down; its angular frequency, $\omega = 190 \text{ s}^{-1}$, is not exactly constant, but changes at a rate given by

$$\frac{d\omega}{dt} = -2.4 \times 10^{-9} \text{ s}^{-2}, \tag{6.68}$$

which corresponds to an increase of about a millisecond in the period every 90 years. If the pulsar is a rotating neutron star, its energy of rotation, $E_{rot} = \frac{1}{2}I\omega^2$, also decreases in accordance with

$$\frac{dE_{rot}}{dt} = I\omega\frac{d\omega}{dt}. \tag{6.69}$$

If we assume a moment of inertia consistent with the estimate given by Eq. (6.67), say $I = 10^{38}$ kg m$^2$, we deduce that the rate of loss of rotational energy of the neutron star at the heart of the Crab Pulsar is $4.6 \times 10^{31}$ W. This energy loss is comparable with the estimated luminosity of the Crab Nebula, $5 \times 10^{31}$ W . It is, therefore, highly likely that the power lost by a rapidly rotating neutron star is the source of the luminosity of the Crab Nebula.

The most likely mechanism for the loss of energy by a rotating neutron star is magnetic dipole radiation; see any good book on electromagnetism, such as Barger and Olsson (1987). If a rotating neutron star has a magnetic dipole inclined at an angle to its axis of rotation, the spinning magnetic dipole radiates electromagnetic radiation; for a star with magnetic dipole $m$ at an angle $\theta$ to an angular velocity $\omega$, energy is radiated at a rate given by

$$P = \frac{2}{3c^3}\left[\frac{\mu_0}{4\pi}\right] m^2\omega^4 \sin^2\theta. \tag{6.70}$$

If this mechanism is responsible for the observed rate of energy loss of the Crab Pulsar, $5 \times 10^{31}$ W, then the neutron star in the Crab Nebula has a magnetic dipole given by

$$m \sin\theta \approx 4 \times 10^{27} \text{ A m}^2. \tag{6.71}$$

It follows that the magnetic field on the surface of the neutron star is approximately

$$B \approx \frac{\mu_0 m}{4\pi R^3} \approx 10^8 \text{ T}, \tag{6.72}$$

if the neutron star has a radius $R$ of about 10 km. This is a huge magnetic field, corresponding to a magnetic energy density $B^2/2\mu_0$ of $4 \times 10^{21}$ J m$^{-3}$. But such a field could arise by the trapping of magnetic flux during stellar collapse to a very compact neutron star. The magnetic flux through any loop moving with a fluid of high conductivity is constant. Thus, the contraction of an iron core of radius 1000 km to a neutron star of radius 10 km could enhance an internal magnetic field by a factor of $10^4$.

There is a compelling historical argument in favour of magnetic dipole radiation as a mechanism for energy loss from pulsars: this mechanism yields an age for the Crab Pulsar that is consistent with the date of the supernova which produced the Crab Nebula, 1054 AD. To show this, we note that Eq. (6.70) indicates that the mechanism predicts a rate of energy loss proportional to $\omega^4$. Thus the rate of change of the rotational energy is given by

$$\frac{dE_{rot}}{dt} = I\omega\frac{d\omega}{dt} \propto \omega^4.$$

Hence the angular velocity of the star satisfies the differential equation

$$\frac{\mathrm{d}\omega}{\mathrm{d}t} = -C\omega^3, \tag{6.73}$$

where $C$ is a constant time which, for the neutron star in the Crab Nebula, can be determined to be $3.5 \times 10^{-16}$ s by using the current values of the angular velocity and acceleration, $\omega = 190\ \mathrm{s}^{-1}$ and $\mathrm{d}\omega/\mathrm{d}t = -2.4 \times 10^{-9}\ \mathrm{s}^{-2}$. If we integrate Eq. (6.73) and set $\omega = \omega_i$ at time $t = 0$, we find

$$t = \frac{1}{2C}\left[\frac{1}{\omega^2} - \frac{1}{\omega_i^2}\right]. \tag{6.74}$$

By substituting the current value for the angular velocity, we conclude that the neutron star in the Crab Nebula has been rotating for a time bounded by

$$t < \frac{1}{2C\omega^2} = 4 \times 10^{10}\ \mathrm{s} = 1253\ \text{years}. \tag{6.75}$$

We note, with satisfaction, that this time is comparable with the historical age of the Crab Nebula, 1993 – 1054 or 939 years. In fact, this model gives the correct historical age if the initial angular velocity of the neutron star was about $400\ \mathrm{s}^{-1}$. However, we should note that the decline in the angular velocity is not the steady decline described by Eq. (6.73). Small, abrupt increases in the angular velocity occur from time to time as the neutron star undergoes internal changes; for example, the Crab Pulsar's 33 millisecond period suddenly decreased by about 3 nanoseconds on 29 August 1989.

Finally, it should be emphasized that the physics underlying pulsar emission mechanisms is extremely complicated. Amongst other things, it involves the interaction of intense, rapidly-rotating magnetic fields with the plasma surrounding neutron stars. As a result, many of the observed features of pulsars cannot be understood using simple models.

### The maximum mass of a neutron star

To a first approximation neutrons play the same supporting role in a neutron star as electrons in a white dwarf. They can also fail to support in similar ways. Just as degenerate electrons are unable to support a white dwarf with a mass above a critical limit, the Chandrasekhar limit, degenerate neutrons are unable to support a neutron star with a mass above a certain value.

The physics underlying the Chandrasekhar limit is clear-cut. As the mass of the white dwarf approaches the limit, the central density increases and the degenerate electrons become increasingly relativistic. At the Chandrasekhar limit, the electrons are ultra-relativistic, the density approaches 'infinity' and the star collapses. A similar phenomenon involving neutrons is expected in a neutron star, but there are a number of important differences. First, the interactions between neutrons are very important at the high densities found in a neutron star. Second, the gravitational fields are very strong and Einstein's theory, not Newton's, should be used to describe the equilibrium of a neutron star under gravity. However, these important differences do not alter the fundamental result that there is a maximum mass for a neutron star. Their main effect is to make the calculation of this maximum mass very difficult.

We emphasize that the actual value for the maximum mass for a neutron star plays a key role in the search for black holes in astronomy. The masses of the stars in a binary system can sometimes be determined from the observed relative motion. If one of the members of the binary is a compact object with a mass greater than the theoretical maximum mass of a neutron star, this object is almost certainly a black hole.

We shall begin our discussion on neutron-star masses by drawing upon our analysis of white dwarfs and finding the neutron-star analogue of the Chandrasekhar mass. We shall crudely ignore the interactions between neutrons and equate the pressure of an ultra-relativistic gas of neutrons to the pressure needed to support a star of mass $M$. In analogy with Eqs. (6.7), (6.8) and (6.18), we arrive at the following expression for the maximum mass supportable by an ideal gas of degenerate neutrons

$$M_{max} \approx 3.1\, M_{\star} = 5.8\, M_{\odot}. \tag{6.76}$$

In effect, we simply set $Y_e = 1$ in the expression for the Chandrasekhar mass for a white dwarf.

As mentioned earlier, the interactions between the neutrons are important in a neutron star. They definitely have a role in determining the maximum value for the mass of a neutron star. These interactions are attractive at inter-nucleon distances around 1.4 fm but repulsive at shorter distances. This would suggest that neutron-star matter becomes harder to compress at high densities. But at high densities, the degenerate neutrons are sufficiently energetic to produce new particles, such as hyperons and pions. If this happens, the pressure due to energetic degenerate neutrons is reduced, but the pressure generated by the new particles is small. Thus, particle production is likely to make neutron-star matter more compressible, an effect which partially offsets the effect of short-range repulsion between neutrons. Overall, the interactions between neutrons tend to increase the theoretical maximum mass of a neutron star.

Einstein's theory of gravity, General Relativity, also plays a crucial part in determining the maximum mass of a neutron star. As indicated in Eq. (6.62), the gravitational binding energy of a massive neutron star is comparable with its rest mass, or more precisely the mass it would have if its constituent particles were at rest and isolated from each other. The estimate of 5.8 $M_{\odot}$ given by Eq. (6.76) corresponds to the rest mass of a star. Gravitational binding implies that the actual mass is considerably smaller. But by far the most important effect of the relativistic effects of gravity is that the attractive nature of gravity is strengthened at very high densities and pressures. This will tend to reduce the theoretical maximum mass of a neutron star.

To appreciate the role of the enhanced effect of gravity at high densities and pressures, we re-examine the equation for the hydrostatic equilibrium of a spherical mass distribution. In Chapter 1 we showed that the internal pressure gradient needed to oppose Newtonian gravity is given by Eq. (1.5), namely

$$\frac{\mathrm{d}P}{\mathrm{d}r} = -\frac{Gm\rho}{r^2}, \tag{6.77}$$

where $m(r)$ is the mass enclosed by a sphere of radius $r$ and $\rho(r)$ is the density at $r$. The corresponding equation in Einstein's theory of gravitation is

$$\frac{\mathrm{d}P}{\mathrm{d}r} = -\frac{Gm\rho}{r^2} \times \frac{(1 + P/\rho c^2)(1 + 4\pi r^3 P/mc^2)}{(1 - 2Gm/rc^2)}. \tag{6.78}$$

Note that the Newtonian equation for hydrostatic equilibrium is recovered if the velocity of light, $c$, tends to infinity.

An essential difference between Newton's and Einstein's theory lies in the source of the gravitational field. In the former it is the mass density, in the latter it is the energy momentum tensor, an entity which depends on the energy density and pressure. As a result, energy and pressure gives rise to gravitational fields in very compact objects. This is illustrated in Eq. (6.78). The terms $m(r)c^2$ and $\rho(r)c^2$ are the energy enclosed by radius $r$ and the energy density at $r$. More importantly, the pressure $P$ occurs on the right-hand side of the equation. This pressure dependence of gravity has a dramatic effect on the stability of neutron stars. It implies that the progressive increase in pressure needed to oppose gravitational collapse is, ultimately, self-defeating because it leads to a strengthening of the gravitational field. Gravity is stronger and collapse is easier.

Any realistic calculation of the properties of neutron stars is based upon the general relativistic equation for hydrostatic equilibrium, and an equation of state for neutron-star matter, $P = P(\rho)$, which takes account of nuclear interactions. Equation (6.78) is integrated starting from $\rho = \rho_c$ at $r = 0$ to the surface at $r = R$ where $\rho = 0$. In this way, one finds a radius $R$ and a mass $M = m(R)$ for a given central density. In particular, the mass for which the star collapses can be found. The first calculation of this kind was by Oppenheimer and Volkoff in 1939. They found that the maximum mass of a star composed of non-interacting neutrons is $0.7\ M_\odot$. This is smaller than the estimate given by Eq. (6.76) because, in General Relativity, the enhanced effect of gravity leads to a collapse at a finite density when the neutrons are becoming relativistic, not when they are ultra-relativistic.

There have been a number of calculations using equations of state corresponding to a range of possible compressibilities for neutron-star matter. The predicted maximum masses range from 1 to 3 $M_\odot$. In fact, the detection of neutron stars with masses around 1.5 $M_\odot$ in binary systems indicates that the compressibility of neutron-star matter is high.

In order to explicitly illustrate the role of the enhanced effect of gravity in neutron stars due to General Relativity, we shall consider an extreme, but very simple, model for matter inside a neutron star. We shall assume matter with a constant density $\rho_0$, which is incompressible at any finite pressure.

We begin by finding the pressure profile inside a star of constant density $\rho_0$ in hydrostatic equilibrium under Newtonian gravity. Integration of Eq. (6.77), implies that the pressure in a such a star is

$$P(r) = G\frac{2\pi}{3}\rho_0^2(R^2 - r^2), \tag{6.79}$$

where the radius $R$ is defined by $P(R) = 0$. We note that the pressure increases quadratically, and at the centre it reaches a value given by

$$P_c = G\frac{2\pi}{3}\rho_0^2 R^2 = \left[\frac{\pi}{6}\right]^{1/3} GM^{2/3}\rho_0^{4/3}, \tag{6.80}$$

where $M = m(R)$ is the mass of the star. We note that this pressure is finite for any finite value of the mass of the star. We conclude, without any surprise, that Newtonian gravity places no restriction on the mass of a star made from incompressible nuclear matter.

The corresponding general relativistic expression for the pressure inside a star of constant density can be found by integrating Eq. (6.78). A little private calculus leads to

$$P = \rho_o c^2 \left[\frac{(1 - 2GMr^2/R^3c^2)^{1/2} - (1 - 2GM/Rc^2)^{1/2}}{3(1 - 2GM/Rc^2)^{1/2} - (1 - 2GMr^2/R^3c^2)^{1/2}}\right]. \tag{6.81}$$

The pressure at the centre of the star is

$$P_c = \rho_0 c^2 \left[\frac{1 - (1 - 2GM/Rc^2)^{1/2}}{3(1 - 2GM/Rc^2)^{1/2} - 1}\right]. \tag{6.82}$$

By considering the denominator of this equation, we see that the central pressure is finite only if

$$\frac{GM}{Rc^2} < \frac{4}{9}. \tag{6.83}$$

This inequality can be rewritten in terms of the mass of the star and its constant density $\rho_0$. But before doing so, we shall express the constant density in terms of the neutron mass as follows,

$$\rho_0 = \frac{3m_n}{4\pi r_n^3} \quad \text{where} \quad r_n = f_n \frac{h}{m_n c}, \tag{6.84}$$

where $f_n$ is a dimensionless length parameter. Note the density of normal nuclear matter, $\rho_{nuc} = 2.3 \times 10^{17}$ kg m$^{-3}$, corresponds to $f_n = 0.9$. We can now rewrite Eq. (6.83) and show that the pressure at the centre of a neutron star of constant density is finite if its mass is smaller than

$$M_{max} = \left[\frac{8\pi f_n}{9}\right]^{3/2} M_\star. \tag{6.85}$$

Yet again we have found that the magnitude of a stellar mass of crucial importance is of the order of $M_\star$, the fundamental stellar mass defined by Eq. (6.59).

We conclude from Eq. (6.85) that even incompressible matter can collapse under gravity. In particular, Einstein's General Relativity imposes an upper limit to the mass of a star made from incompressible nuclear matter; if this mass is exceeded, the internal pressure needed to support the star becomes infinite. This maximum mass depends on the value of the assumed constant density. For a star made up of matter with a density equal to double the normal density of nuclear matter, the length parameter $f_n$ is 0.7, and the maximum mass is 2.7 $M_\star$ or 5 $M_\odot$.

In conclusion, we have two crude estimates for the maximum mass of a neutron star: one based on the stability of a compressible ideal gas under Newtonian gravity, Eq. (6.76), and one based on the stability of incompressible, constant-density nuclear matter under general relativistic gravity, Eq. (6.85). The usefulness of these estimates lies not in their numerical values, but in the physical ideas underlying their derivation. Realistic calculations must take into account the compressibility of neutron-star matter and General Relativity. The consensus reached by these calculations is that the maximum possible mass of a neutron star is probably smaller than 3 $M_\odot$ and definitely less than 5 $M_\odot$.

## 6.4 BLACK HOLES

We now turn to the fate of a collapsing stellar core which is too massive to end its life as a neutron star. As the collapse proceeds the gravitational field becomes stronger and stronger, and the internal pressure becomes larger and larger. But the source of the gravitational field in General Relativity is the energy density and the pressure. Hence the increase in pressure accelerates the final stages of collapse. According to General Relativity the star enters a region of space-time called a black hole. Nothing can halt the collapse. Nothing can escape, not even light. And nothing is left of the collapsed stellar core apart from an extremely strong gravitational field.

Gravitational collapse, the driving mechanism of stellar evolution, has progressed to its ultimate end, infinite compression.

In the opening paragraph we attributed the collapse to a black hole to a progressively increasing force of gravitational attraction. But this description improperly treats space and time as two separate concepts. It is more accurate to describe a black hole in terms of a distortion of the unified concept of space-time. In General Relativity, gravity is not a force, but a distortion of the geometrical properties of space-time due to the presence of matter and radiation. The sun only produces a slight 'dent' in space-time, but a collapsed core of a massive star can produce a 'hole'. Nothing can escape from this hole because there are no outward paths in this distorted region of space-time; every path is towards the centre of the hole. It is a hole of no return.

The size of a black hole depends on the mass of the collapsed object. It is the Schwarzschild radius[1]

$$R_{sch} = \frac{2GM}{c^2}. \tag{6.86}$$

For a collapsed mass equal to 10 $M_\odot$, the Schwarzschild radius is 30 km. The Schwarzschild radius marks the boundary of the one-way surface of the black hole. This surface is not made of anything. It encloses an unobservable region of space in which all motion is towards the centre.

A black hole is formed when the radius of a collapsing star reaches the Schwarzschild radius. If a distant observer could view the collapse s/he would see the star frozen at this radius, because time in the intense gravitational field on the surface of the star appears to grind to a halt. This gravitational field leads to a gravitational red shift which 'extinguishes' the star as the Schwarzschild radius is approached. In fact, the fractional change in wavelength of radiation escaping from the surface of a star is given by Eq. (6.27), which, when rewritten in terms of the radius of the star and the Schwarzschild radius, becomes

$$\frac{\Delta\lambda}{\lambda} = \left[1 - \frac{R_{sch}}{R}\right]^{-1/2} - 1. \tag{6.87}$$

This indicates that the red shift tends to infinity as $R$ approaches $R_{sch}$. The frequency of the radiation tends to zero so that the energy escaping from the star also tends to zero. Indeed, the luminosity decreases exponentially in accordance with

$$L \propto \exp\left[-\frac{t}{\tau}\right] \quad \text{where} \quad \tau = \frac{R_{sch}}{c}. \tag{6.88}$$

For a collapsed mass equal to 10 $M_\odot$, $\tau = 10^{-4}$ s.

[1] By coincidence the correct expression for the Schwarzschild radius can be obtained by setting the Newtonian escape velocity of a particle from an object of mass $M$ and radius $R$ equal to $c$.

However, the star only appears to freeze and fade into darkness to a distant observer. To an observer within the Schwarzschild radius the star is still active and lively. Indeed, such an observer would find out what happens to all the quarks, electrons, neutrinos and photons inside the black hole. Unfortunately, s/he will be cut off from the rest of the universe and will be unable to share the knowledge.

The detection of a black hole, an object whose only manifestation is an intense gravitational field, is not an easy task. Any evidence for its existence must be circumstantial because it cannot be seen. But the immense gravitational attraction of a black hole can reveal its presence.

For example, when gaseous matter is pulled towards a black hole it acquires kinetic energy and becomes very hot. The resulting temperature of this gas and the nature of its radiation, before it is hidden for ever within the Schwarzschild radius, are a measure of the strength of the gravitational field it is entering. In particular, the accretion of matter onto a black hole is expected to be accompanied by X-ray radiation. However, similar X-ray radiation can also be produced by gaseous matter entering the strong gravitational field of a neutron star. But the presence of a neutron star can be ruled out if the mass of the compact object involved exceeds the maximum possible mass of a neutron star. The detection of any compact object with a mass greater than this limit is, by default, a black hole.

The currently favoured method for detecting black holes is based upon the observation of compact X-ray sources, binary systems consisting of a visible ordinary star and an invisible compact object. X-rays are produced by mass flowing from the ordinary star into the strong gravitational field of the invisible compact object. Information on the relative motion of the binary system can be deduced from the spectrum of radiation from the ordinary star, and this information can be used to set limits on the mass of the invisible, compact object. If this mass is greater than 3 $M_\odot$, the compact object is probably a black hole. If its mass is greater than 5 $M_\odot$, it almost certainly is a black hole.

The first and most famous candidate for a black hole is the compact object in the binary X-ray source Cygnus X-1. The inferred relative motion of this binary system implies the presence of a compact object with a mass definitely greater than 3.4 $M_\odot$. Another candidate for a black hole is provided by the X-ray system V404 Cygni, where there is compact object with a mass greater than 6.3 $M_\odot$. In fact, evidence for the existence of stellar black holes is very convincing (see, for example, Shapiro and Teukolsky (1989) and Casares, Charles and Naylor (1992)).

The possibility that the evolution of a star could lead to the formation of a black hole was first recognized in the 1930's, soon after Chandrasekhar's discovery of a maximum value for the mass of a white dwarf. The existence of this maximum implied that a massive stellar core could collapse into a region of space in which gravity was overwhelming. Many astrophysicists found this outcome for stellar evolution unacceptable, if not absurd. Eddington, as usual, made his view very clear when he wrote in 1935:

*The star apparently has to go on radiating and radiating and contracting and contracting until, I suppose, it gets down to a few kilometres radius when gravity becomes strong enough to hold the radiation and the star at last can find peace... I think that there should be a law of Nature to prevent the star from behaving in this absurd way.*

However, the current belief is that a black hole, like a white dwarf and a neutron star, is a respectable end-point for stellar evolution. It is a belief based on firm theoretical foundations and supported by evidence from observational astronomy.

Gravity is the driving force for stellar evolution. It leads to the formation of a star and to temperatures which make thermonuclear fusion possible. The energy released by fusion only serves to delay the gravitational contraction of the matter inside the star. The end-point may be a white dwarf or a neutron star, stars in which cold matter resists the force of gravity. An alternative end-point is a black hole in which gravity is completely triumphant. This outcome is neat and tidy: nothing is left of the collapsed matter apart from an intense gravitational field.

## SUMMARY

### White dwarfs

- To a first approximation a white dwarf is a star supported by the pressure of an ideal gas of degenerate electrons.
- The degenerate electrons at the centre of a white dwarf with a low mass are non-relativistic, and the central density increases with the mass $M$ in accordance with

$$\rho_c \approx \frac{3.1}{Y_e^5}\left[\frac{M}{M_\star}\right]^2 \frac{m_{\mathrm{H}}}{(h/m_e c)^3}, \tag{6.4}$$

  where $Y_e$ is the number of electrons per nucleon and $M_\star$ is the fundamental stellar mass defined by Eq. (5.63). As the mass increases, the electrons become relativistic and the density increases more rapidly. As the mass approaches the Chandrasekhar limit, the electrons become ultra-relativistic and the central density tends to infinity, as shown in Fig. 6.1. In other words, the star collapses.
- The Chandrasekhar limit is the mass of the white dwarf whose central density tends to infinity. As such, it represents the maximum possible mass for a white dwarf. The pressure–density relation (6.6) leads to an estimate for $M_{CH}$ given by Eq. (6.9). A more accurate estimate, based on a polytrope model, is given by

$$M_{CH} \approx 3.1\, Y_e^2\, M_\star, \tag{6.18}$$

  which corresponds to a mass of about 1.4 solar masses.

- The radius of a white dwarf is a decreasing function of its mass. For white dwarfs of low mass, the approximate relation between radius and mass is

$$R \approx 0.77\, Y_e^{5/3} \left[\frac{M_\star}{M}\right]^{1/3} \alpha_G^{-1/2} \frac{h}{m_e c}. \tag{6.20}$$

This implies that the characteristic size of a white dwarf is

$$\alpha_G^{-1/2} \frac{h}{m_e c} \approx 3 \times 10^7 \text{ m}. \tag{6.21}$$

If we use the sun as a standard for mass and size, and if $Y_e = 0.5$, then

$$R \approx \frac{R_\odot}{74} \left[\frac{M_\odot}{M}\right]^{1/3}. \tag{6.23}$$

This mass–radius relation can be used to derive expressions for the luminosity and gravitational red shift of radiation from a white dwarf in terms of its mass; see Eqs. (6.25) and (6.28).

## Stellar collapse

- Nuclear photodisintegration and electron capture are two possible mechanisms for the absorption of energy which could drive the iron core of a star into uncontrolled collapse.
- The energy of formation of a neutron star, essentially the gravitational binding energy of a neutron star,

$$E_B \approx \frac{GM^2}{R} = 3 \times 10^{46} \left[\frac{M}{M_\odot}\right]^2 \left[\frac{10 \text{ km}}{R}\right] \text{ J}, \tag{6.46}$$

is an order of magnitude larger than the energy absorbed by nuclear photodisintegration or by electron capture. The bulk of this energy is emitted in the form of neutrino radiation; see Fig. 6.3.

## Neutron stars

- If we assume that Newtonian gravitation in a neutron star is opposed by the pressure of an ideal gas of degenerate, non-relativistic neutrons, then the radius of a star of mass $M$ is given by

$$R \approx 0.77 \left[\frac{M_\star}{M}\right]^{1/3} \alpha_G^{-1/2} \frac{h}{m_n c}. \tag{6.60}$$

This implies that the characteristic size of a neutron star is

$$\alpha_G^{-1/2}\frac{h}{m_n c} \approx 17 \text{ km}. \tag{6.61}$$

- Estimates for the typical mass and radius of a neutron star lend support to the hypothesis that pulsars are rapidly rotating neutron stars.
- Neutron stars cannot have a mass greater than a certain critical limit, the analogue of the Chandrasekhar limit for white dwarfs. However, the gravitational fields are strong and General Relativity must be used. The exact value of the maximum possible mass of a neutron star is difficult to calculate because of the uncertainty in the compressibility of neutron-star matter at high densities. It is probably around 3 $M_\odot$, and almost certainly below 5 $M_\odot$.

### Black holes

- If a collapsed stellar core has a mass greater than the maximum mass of a neutron star, it will undergo complete collapse and form a black hole.
- Any method for detecting a black hole depends on observing the effects of its intense gravitational field. The observation of some compact X-ray sources indicates the presence of intense gravitational fields due to compact objects which are too massive to be neutron stars. These objects, by default, are thought to be black holes.

## PROBLEMS 6

6.1 According to Eq. (6.4), the central density of a body supported by degenerate electrons goes to zero as the mass of the body goes to zero. This unphysical result arises from the neglect of electromagnetic interactions between electrons and ions. In fact, as the pressure falls, the density tends to a value corresponding to ordinary, uncompressed atomic matter. Because the size of an atom is of the order of the Bohr radius, this density is approximately given by

$$\rho_{atomic} = \frac{m_{\mathrm{H}}}{a_B^3},$$

where $a_B$, the Bohr radius, can be written as

$$a_B = \frac{1}{\alpha_{EM}}\frac{\hbar}{m_e c}.$$

The constant $\alpha_{EM}$ is a dimensionless measure of the strength of the electromagnetic interaction, the fine structure constant $\alpha_{EM} = e^2/(4\ \pi\epsilon_0\,\hbar\ c) = 1/137$. Show that the central density of a body supported by degenerate electrons becomes comparable with normal atomic densities when the mass of the body is comparable with

$$M_P = \left[\frac{\alpha_{EM}}{\alpha_G}\right]^{3/2} M_{\mathrm{H}} = \alpha_{EM}^{3/2}\, M_\star = 0.001\, M_\odot.$$

Note we can roughly identify this mass with the maximum mass of a body containing ordinary atomic matter, in other words the maximum mass of a body like a planet. Indeed, the mass of Jupiter is 0.00095 $M_\odot$.

6.2 The energy needed to dissociate an $^4$He nucleus into two neutrons and two protons is $Q$= 28.3 MeV. Derive an expression relating the numbers of $^4$He nuclei, neutrons and protons coexisting at a temperature $T$ in an equilibrium set up by the reactions

$$\gamma + ^4\mathrm{He} \rightleftharpoons 2n + 2p.$$

Calculate the temperature for 50% dissociation when the density is $10^{12}$ kg m$^{-3}$.

6.3 Assume that a hot, bloated neutron star emits thermal neutrino radiation from a surface of radius $R$ at an effective temperature equal to $T_E$. Assume that three types of massless, or nearly massless, neutrinos, $\nu_e, \nu_\mu, \nu_\tau$ and their antiparticles, are emitted in equal numbers, in thermal equilibrium with zero chemical potential. Show that the luminosity is given by

$$L_\nu = \frac{21}{8} \sigma T_E^4 4\pi R^2,$$

where $\sigma$ is Stefan's constant. Find an expression for the average energy for a neutrino in this radiation. [Hint: Look back at Chapter 2 and reconsider problem **2.5**.]

6.4 The outward expulsion of the outer layers of a massive star by a shock wave generated by core rebound is the most promising mechanism for generating a supernova from gravitational collapse. A possible alternative mechanism involves neutrinos. Neutrino radiation from the collapsed core could transmit outward momentum and cause an expulsion. By reconsidering problem **3.3** at the end of Chapter 3, show that this mechanism could be effective only if the neutrino luminosity exceeds a value given by

$$L_\nu > \frac{4\pi c G M}{\kappa_\nu},$$

where $M$ is the mass of the collapsed core and $\kappa_\nu$ is the neutrino opacity. By noting that the neutrino opacity is of the order of $10^{-18}$ m$^2$ kg$^{-1}$, show that the expected neutrino luminosity of around $10^{45}$ W is insufficient to cause an expulsion.

6.5 The detection of neutrinos (mostly $\overline{\nu}_e$'s in fact) from the supernova SN1987A at a distance of 50 kpc from the Earth provided valuable information on the maximum possible mass of the electron neutrino. Write down a general expression for the velocity of a neutrino of mass $m$ and energy $E$ as a fraction of the velocity of light, and confirm that more energetic neutrinos move faster and arrive earlier at the Earth. Show that, if the mass of the neutrino is 30 eV $c^{-2}$, then a 10 second spread in the arrival time at the Earth is expected for neutrinos with energy between 10 and 15 MeV.

6.6 Consider an ideal degenerate gas of electrons, protons and neutrons, and the equilibrium established by the reactions (6.52). Assume equal numbers of electrons and protons and assume that the density is so high that all the degenerate particles are ultra-relativistic. Show that the number densities of the particles are in the ratio

$$n_e : n_p : n_n = 1 : 1 : 8.$$

6.7 Estimate the maximum angular velocity of rotation of a typical white dwarf.

6.8 The ratio of the Schwarzschild radius to the actual radius of a body is the crucial parameter for assessing the importance of General Relativity. Show, that for a main sequence star, like the sun, with a typical interior temperature $T_I$, this ratio is approximately given by

$$\frac{R_{sch}}{R} \approx \frac{kT_I}{m_{\mathrm{H}}c^2}.$$

Show that for a white dwarf

$$\frac{R_{sch}}{R} \approx \frac{m_e}{m_{\mathrm{H}}},$$

and for a neutron star

$$\frac{R_{sch}}{R} \approx 1.$$

6.9 The Crab Pulsar is a rotating neutron star formed by a supernova in 1054 AD. At present it has an angular velocity and an angular acceleration given by:

$$\omega = 190 \text{ s}^{-1} \quad \text{and} \quad \frac{\mathrm{d}\omega}{\mathrm{d}t} = -2.4 \times 10^{-9} \text{ s}^{-2}.$$

If gravitational radiation were responsible for the Crab slowdown, the rate of loss of rotational energy would be proportional to $\omega^6$. Use this model to derive an expression for the time dependence of $\omega$. Show that this model predicts an age which is less than the actual age of the pulsar.

6.10 Consider a compact X-ray source consisting of an ordinary star and a neutron star. Assume that a X-ray luminosity of $10^{31}$ W is powered by the release of gravitational energy due to the accretion of mass onto the surface of a neutron star.

1. Use Eq. (6.62) to show that the luminosity is consistent with a mass accretion rate of about $10^{-8}\ M_{\odot}$ per year.
2. Show that, if the source of this luminosity is a black body whose size is comparable with a typical neutron star, then the effective surface temperature of the source is such that the radiation is indeed in the X-ray region of the electromagnetic spectrum.

# Hints to selected problems

## CHAPTER 1

1.1 Make use of Eqs. (1.6), (1.7) and (1.5).

1.2 Use the equation before Eq. (1.19) to estimate the minimum mass that could condense under gravity at the temperature and density given.

1.4 Bear in mind that the luminosity of a star of mass $M$ is proportional to $M^{\alpha}$ with $\alpha$ between 3 and 3.5; see Fig. 1.4.

1.6 Find the energy flux from the sun at a distance of 10 pc. Such a sun would appear as a star of magnitude 4.72. Use Eq. (1.40) to compare the energy flux received from stars of magnitude 6 and 4.72.

1.7 Let

$$F(r) = \left[P(r) + \frac{Gm(r)^2}{8\pi r^4}\right] \quad \text{and show that} \quad \frac{dF}{dr} < 0.$$

The first lower bound on $P_c$ is given by the condition $F(0) > F(R)$.
The second lower bound and the upper bound on $P_c$ can be obtained by noting that

$$m(r) > \frac{4\pi}{3}\langle\rho\rangle r^3, \; m(r) < \frac{4\pi}{3}\rho_c r^3,$$

and using

$$\frac{dF}{dr} + \frac{Gm(r)^2}{2\pi r^5} = 0.$$

## CHAPTER 2

2.2 Make use of Eq. (2.36).

2.4 Derive the Saha equation corresponding to $\gamma + H_2 \rightleftharpoons H + H$,

$$\frac{n(H)n(H)}{n(H_2)} = \left[\frac{\pi m_H kT}{h^3}\right]^3 \exp\left[-(4.48\text{eV})/kT\right],$$

and use $P = [n(H) + n(H_2)\ ]\ kT$. Impose the condition for 50% dissociation, $n(H_2) = 2\ n(H)$.

2.5 The calculation closely follows that leading to Eqs. (2.42) and (2.43). The substitution of massless fermions for bosons leads to integrals with a denominator $e^x + 1$ instead of $e^x - 1$. Expand each integral as a series, rearrange the series and express it in terms of the Riemann Zeta Function. For example, the integral occurring in the calculation of density of fermions is

$$\int_0^\infty \frac{x^2 dx}{e^x + 1} = 2\left[\frac{1}{1^3} - \frac{1}{2^3} + \frac{1}{3^3} - \cdots\right] = 2\left[\left(\frac{1}{1^3} + \frac{1}{2^3} + \cdots\right) - 2\left(\frac{1}{2^3} + \frac{1}{4^3} + \cdots\right)\right].$$

Hence

$$\int_0^\infty \frac{x^2 dx}{e^x + 1} = 2\left[\zeta(3) - \frac{1}{4}\zeta(3)\right] = \frac{3}{2}\zeta(3).$$

2.6 Consider the equilibrium established by

$$\gamma + \gamma \rightleftharpoons e^+ + e^-.$$

Use Eq. (2.21) for the chemical potential for the dilute gas of positrons and the Fermi energy for the chemical potential for the dense gas of degenerate electrons.

## CHAPTER 3

3.1 The opacity is $\kappa = 1/\rho\bar{l}$ and the frequency averaged mean free path is given by Eq. (3.14).

3.3 The radiation pressure is $P_r = a\ T^4/3$, hence by Eq. (3.28)

$$\frac{dP_r}{dr} = -\frac{\rho\kappa}{c}\frac{L}{4\pi r^2}.$$

Equate this pressure gradient to the gravitational force on unit volume of matter near to the surface of the star.

3.5 Use Eq. (3.37) to relate the fractional differences in the temperature and pressure at nearby points in the envelope of the white dwarf, and compare with the condition for convection, Eq. (3.22).

## CHAPTER 4

4.1 Make use of Eq. (4.12).

4.3 Note that two proton–proton fusions are needed to produce a $^4He$ nucleus via branch I, but only one is needed if a $^4He$ nucleus is produced via branch II.

4.4 Make use of Eq. (4.29).

4.5 Use Eq. (4.29) again.

4.6 Make use of Eqs. (4.65) and (4.66).

4.7 Use Eq. (2.46). In making the estimate, note that $kT$ is small compared with 9.98 MeV.

4.8 Equate the energy produced by fusion in time $\tau$ to the energy needed to heat the gas to a temperature $T_{ign}$. The former can be found from Eq. (4.19). The latter is approximately the kinetic energy of the particles in an ionized gas at $T_{ign}$; confirm this by estimating the energy needed to ionize the gas.

## CHAPTER 5

5.1 Relate $E_{GR}$ to an integral involving $P(r)$ and evaluate the integral by integration by parts. For the last part use Eq. (5.32).

5.2 The star contracts until $L_{fus}$ reaches $L_{rad}$. At this stage

$$\frac{M^6}{R^7} \propto \frac{M^{5.5}}{R^{0.5}}.$$

For the last part use $L \propto R^2 T_E^4$.

5.4 If

$$\beta = \frac{P_g}{P_c} = \frac{\rho_c k T_c}{\bar{m} P_c} \quad \text{and} \quad (1-\beta) = \frac{P_r}{P_c} = \frac{a T_c^4}{3 P_c},$$

then

$$\frac{(1-\beta)}{\beta^4} = \frac{a}{3}\left[\frac{\bar{m}}{k}\right]^4 \frac{P_c^3}{\rho_c^4}.$$

The inequality for $P_c$ yields

$$\frac{(1-\beta)}{\beta^4} < \frac{a}{3}\left[\frac{\bar{m}}{k}\right]^4 \frac{\pi}{6} G^3 M^2.$$

## CHAPTER 6

6.3 This problem is a minor variation on problem 2.5. It involves relating the power radiated by a neutrino black body radiator to the energy density in a neutrino gas; the discussion leading to Eqs. (2.45) and (2.46) may be helpful. Note also that, unlike photons or electrons, each neutrino has only one possible polarization.

6.5 The velocity of a neutrino of mass $m$ with an energy $E$ is given by

$$\frac{v}{c} = 1 - \frac{m^2c^4}{2E^2} \quad \text{if} \quad E >> mc^2.$$

A distance of 50 kpc is about $1.6 \times 10^5$ light years.

6.6 Show, in analogy with Eq. (6.56), that the concentrations of degenerate, ultra-relativistic electrons, protons and neutrons in a very dense gas are related by

$$n_e^{1/3} + n_p^{1/3} = n_n^{1/3}.$$

6.8 Make use of Eqs. (1.31), (6.20), and (6.60).

6.9 Make the appropriate modifications to the analysis leading to Eqs. (6.73) to (6.75).

6.10 1. Assume that the luminosity due to accretion of mass by the neutron star equals the rate of loss of gravitational potential energy.
2. Use the definition of the effective surface temperature, Eq. (1.43), and assume a reasonable value for the radius of the neutron star.

# Bibliography

This bibliography lists books and articles which have been explicitly referred to in the text, and other books and articles which will be of interest to the reader.


Baade, W. and Zwicky, F. (1934) Supernovae and cosmic rays, *Physical Review*, Vol. 45, 138.

Bahcall, J. N. (1989) *Neutrino Astrophysics*, Cambridge University Press, Cambridge.

Bahcall, J. N. (1990) The solar neutrino problem, *Scientific American*, May 1990.

Barger, V. D. and Olsson, M. G. (1987) *Classical Electricity and Magnetism*, Allyn and Bacon, Inc., Boston.

Bernstein, J., Brown, L. S. and Feinberg, G. (1989) Cosmological helium production simplified, *Reviews of Modern Physics*, Vol. 61, No. 1, page 25.

Bethe, H. A. (1939) Energy production in stars, *Physical Review*, Vol. 55, 434.

Bethe, H. A. and Brown, G. E. (1985) Supernovae, *Scientific American*, April 1985.

Bowers, R. and Deeming, T. (1984) *Astrophysics I, Stars*, Jones and Bartlett Publishers, Boston.

Burrows, A. (1989) Neutrinos from supernovae, an article in *Supernovae*, edited by A. G. Petschek, Springer-Verlag.

Burrows, A. (1991) The SN1987 neutrino signal and the future, an article in *Supernovae*, edited by S. E. Woosley, Springer-Verlag.

Casares, J., Charles, P. A. and Naylor, T. (1992) A 6.5-day periodicity in the recurrent nova V404 Cygni implying the presence of a black hole, *Nature*, Vol. 355, page 614.

Chandrasekhar, S. (1931) The maximum mass of ideal white dwarfs, *Astrophysical Journal*, Vol. 74, page 81.

Chandrasekhar, S. (1934) Stellar configurations with degenerate cores, *Observatory*, Vol. 57, page 373.

Clayton, D. D. (1983) *Principles of Stellar Evolution and Nucleosynthesis*, University of Chicago Press, Chicago.
Clayton, D. D. (1986) Solar structure without computers, *American Journal of Physics*, Vol. 54, page 354.
Eddington, A. S. (1935) Minutes of the Royal Astronomical Society, *Observatory*, Vol. 58, page 37.
Fredrick, L. W. (1989) Astronomy and astrophysics, Chapter 3 of *Physics Vade Mecum*, edited by H. L. Anderson, American Institute of Physics, New York.
Goldberg, H. and Scadron, M. D. (1981) *Physics of Stellar Evolution and Cosmology*, Gordon and Breach Science Publishers, Inc., New York.
Hewish, A., Bell, S. J., Pilkington, J. D. H., Scott, P. F. and Collins, R. A. (1968) Observation of rapidly pulsating radio source, *Nature*, Vol. 217, page 709.
Hoyle, F., Dunbar, D. N. F., Wenzel, W. A. and Whaling, W. (1953) A state of $^{12}C$ predicted from astrophysical evidence, *Physical Review*, Vol. 92, page 1095.
IMB Collaboration (1987) Bionta *et al.*, *Phys. Rev. Lett.*, Vol. 58, page 1494.
Kamiokande II Collaboration (1987) Hirata *et al.*, *Phys. Rev. Lett.*, Vol. 58, page 1490.
Kittel, C and Kroemer, H. (1980) *Thermal Physics*, W. H. Freeman and Company, San Francisco.
Nauenberg, M. and Weisskopf, V. F. (1978) Why does the sun shine? *American Journal of Physics*, Vol. 46, page 23.
Oppenheimer, J. R. and Volkoff, G. M. (1939) On massive neutron cores, *Physical Review*, Vol. 55, page 374.
Rolfs, C. E. and Rodney, W. S. (1988) *Cauldrons in the Cosmos: Nuclear Astrophysics*, University of Chicago Press, Chicago.
Salpeter, E. E. (1952) Nuclear reactions without hydrogen *Astrophysical Journal*, Vol. 115, page 327.
Salpeter, E. E. (1966) Dimensionless ratios and stellar structure, an article in *Perspectives in Modern Physics, Essays in Honor of Hans A. Bethe*, edited by R. Marshak, Interscience, New York.
Sexl, R. and Sexl, H. (1979) *White Dwarfs-Black Holes: An Introduction to Relativistic Astrophysics*, Academic Press, New York.
Shapiro, S. L. and Teukolsky, S. A. (1983) *Black Holes, White Dwarfs, and Neutron Stars*, John Wiley & Sons, New York.
Shkolovskii, I. S. (1978) *Stars, Their Birth, Life and Death*, W. H. Freeman and Company, San Francisco.
Tayler, R. J. (1972) *The Stars, Their Structure and Evolution* Springer-Verlag, New York.
Weisskopf, V. F. (1975) Of Atoms, Mountains, and Stars: A Study in Qualitative Physics, *Science*, Vol. 187, page 605.

# Index